Lean Auto Body

The Lean Implementation Guide for the Auto Collision Repair Industry

Lean Auto Body

The Lean Implementation Guide for the Auto Collision Repair Industry

CRC Press
Taylor & Francis Group
Boca Raton London New York

CRC Press is an imprint of the
Taylor & Francis Group, an **informa** business
A PRODUCTIVITY PRESS BOOK

Chris A. Ortiz

Originally published as *Lean Auto Body The Lean Implementation Guide to the Auto Collision Repair Industry*,

Published 2023 by CRC Press

Taylor & Francis Group
6000 Broken Sound Parkway NW, Suite 300
Boca Raton, FL 33487-2742

CRC Press is an imprint of Taylor & Francis Group, an Informa business

ISBN-13: 978-1-926537-17-7 (pbk)

Visit the Taylor & Francis Web site at
http://www.taylorandfrancis.com

and the CRC Press Web site at
http://www.crcpress.com

Library of Congress Control Number: 2010920893

Library of Congress Cataloging-in-Publication Data
Ortiz, Chris
Lean Auto Body: The Lean Implementation Guide to the Auto Collision Repair Industry

Distributed by Productivity Press, an imprint of CRC Press

Table of Contents

About the Author

Chris Ortiz is the president and executive consultant of Kaizen Assembly, Inc., a lean manufacturing training and implementation company. Chris has been practicing and teaching lean for nearly 12 years. He is the author of four books (listed below), and has written dozens of training manuals, as well as industry articles for national magazines and publications. His work has been featured in Industrial Engineer, Assembly Magazine, Seattle Times, Industrial Management, and various other news journals and newspapers.

Chris and Kaizen Assembly were featured on the national television show *Inside Business Report*, which aired on CNBC and Headline News, and he travels around the country speaking at engineering and manufacturing trade shows. Along with Kaizen Assembly's growing manufacturing client list, they have been helping many auto body and auto repair shops implement lean and witness the amazing results that can be achieved from it.

Chris is one of a very small list of lean practitioners who understands the auto body industry and how lean is applied to its unique processes.

Chris and his company can be contacted by going to **www.kaizenassembly.com** or by emailing him directly at **chrisortiz@kaizenassembly.com**.

Books by the Author

Kaizen Assembly: Designing, Constructing, and Managing a Lean Assembly Line (CRC Press 2006)

Lessons from a Lean Consultant: Avoiding Lean Implementation Failure on the Shop Floor (Prentice Hall 2008)

Kaizen and Kaizen Event Implementation (Prentice Hall 2009)

Visual Controls: Applying Visual Management in the Factory (Taylor and Francis 2010)

Introduction

Main street perceptions of the lean philosophy are that its concepts can only be applied to the manufacturing industry, and that the best candidates for its implementation are companies producing only one type of product. My experience with the auto collision repair industry is similar in that regard. Often the argument is that body shops do not have repetitive work. The diversity which exists in the type and quantity of automobiles, as well as the complexity of the damage, makes it nearly impossible, some believe, to successfully apply lean manufacturing techniques. However, quite the opposite is true. Think about your own shop. The level of repairs needed from one automobile to another is not infinite. If it were, repairs would never be complete. Whenever a process is complex and varied, lean principles can and should be applied. Also, most auto body shops engage in the same type of common auto repair processes, such as estimating, teardown, prep, painting, drying/curing, re-assembly, and possibly detailing or cleaning. Of course, the amount of time spent on each process is contingent on the level of damage, as well as parts lead time from dealers and auto parts supply vendors.

Lean is about waste reduction, where waste is defined as any non value-added work within each operational process. This includes any unnecessary administrative procedures, which can cause a multitude of problems that ultimately increase costs. I will discuss these wastes in greater detail in Section One. For now, let's focus on the main issues to be addressed.

Ask yourself the following questions:

- Do your technicians spend time sifting through tool chests, looking for tools?
- When a technician cannot locate a necessary tool, do they leave their work area and borrow, or "steal", the tool from a co-worker or another work area?
- Are all your estimates 100% error-free or does vital information that has been forwarded by the insurance companies contain administrative mistakes, which will cause issues for your technicians?
- Do you have automobiles sitting idle for hours, waiting on parts to arrive?
- Do those same automobiles sit outside, waiting for a work area in the shop to become available?
- Are your shop supplies disorganized or hidden behind cabinets, causing technicians to spend time hunting for the appropriate item to perform the work?
- Does it take longer than necessary to set up your paint operation?
- Are excessive amounts of supplies and parts sitting around in the parts room, taking up space and tying up money?
- Do people meander around your shop, continuously looking for things?
- Do your technicians receive the wrong parts, after waiting days for the correct one to arrive?
- Do people wander around the shop asking questions about pick up dates or which job to work on next?

The list of questions you could ask are endless, but the answers are critical when attempting to run an effective and efficient repair shop. Right now the single most important question you should ask yourself is how many of these time-wasting, inefficient, profit-robbing activities occur in your operation, and how often?

The auto body repair industry is "ripe" for lean and the industry, as a whole, is one of the most competitive I have ever observed. Auto body repair cannot be outsourced to other countries like China or India, which allows for a lot of local competition. In the county I reside in, more than 40 body shops exist, and every time an accident occurs, these same 40 shops compete for the repair work. Car owners and insurance companies can select from numerous vendors, which creates a highly competitive market. If you want some, or all of that business, you need to make sure that your shop performs to the best of its ability—at all times.

A body shop is a production factory for repairs and it is critical that you start thinking about it that way. Lean helps eliminate hidden costs, which often go unnoticed; hidden costs are wastes that drive down productivity and, ultimately, squander profit. The auto body industry also operates under numerous fixed costs, which are directly related to the insurance and estimating process, and are outside of your control. Wouldn't it be nice to increase the number of repairs your shop can perform, without raising those fixed costs? Or, to reduce those variable costs, which you can control, and be able to repair even more vehicles? If your answer to both questions is 'Yes', this guide is exactly what you need to help you begin your lean auto body journey.

Lean Auto Body is a how-to guide that provides a step-by-step approach to implementing lean in your shop. This guide was specifically designed for the auto body repair industry, and therefore will only discuss the lean tools and techniques that make sense for a body shop. The success of our auto body and repair clients has proven how powerful and effective lean can be toward eliminating costs and increasing profitability. I would like to show you how lean can help your business thrive in today's competitive economy. You will be embarking on a journey that will transform your shop into a robust, efficient, and highly competitive enterprise. It all starts with learning about how lean can help.

Read this guide and make it happen.

Chris Ortiz, President

Kaizen Assembly, Inc.

Section One – The Eight Wastes

Waste exists at every level of a body shop. Waste is defined as any non value-added work that occurs during any phase of the repair process that your customer has no interest in. Essentially, two types of work are performed in any shop: value-added work and non value-added work.

Value-Added Work

Value-added work is defined as any work performed which the customer cares about.

Value-added work includes:

- Tearing down a car
- Prepping panels for paint
- Applying filler
- Painting the car
- Installing parts and replacing vital fluids

Non Value-Added Work

Non value-added work is defined as any work performed which the customer does not care about.

Non value-added work includes:

- Searching for tools and supplies
- Setting up the paint booth
- Traveling to and from the office
- Idle cars waiting for parts
- Transporting cars around the shop
- Making mistakes during repairs
- Cars returned by the customers
- Extended sanding and buffing time

I am not implying that non value-added work will be completely eliminated, however, the primary goal of a lean journey is continuous improvement and waste reduction. Why should your technicians walk fifty feet to retrieve a tool from a cabinet, only to find that it doesn't work properly? Why should they dig through piles of assorted tools and gadgets, wasting valuable work time? Wouldn't it be more cost effective to have all necessary tools organized, in good working order, and within the reasonable reach of your technicians?

The Eight Wastes

The eight wastes of an auto body repair shop are:

- Overproduction
- Extra processing
- Transportation
- Motion
- Waiting
- Defects/Rejects
- Inventory
- Non-Utilized Talents

Overproduction

Overproduction is the act of producing or "doing" more than required, faster than necessary, and before it is needed. Do your technicians begin working on repairs just to stay busy, before the necessary parts or information are available? Are cars with later commit dates being repaired before cars that require faster turn-around? Are estimates being processed in the wrong order, before your shop actually needs them?

Extra Processing

Extra Processing is redundant effort and extra steps, which prolong lead times. Are there extra checks and balances in the estimating process that could be removed? Re-checking and re-verifying, due to lack of faith in the quality of information or preparation, can contribute to an increase in overtime. As well, technicians can over process or over perform during value-added work such as grinding, sanding and buffing. This type of extra processing occurs when the person conducting the work is unable to identify when the work has been completed.

To prove this point, give three of your technicians the same identical body panel, requiring an equal amount of buffing. Provide them with the same buffing tool and material. Then, say go! More than likely, each technician will finish at a different time, due to their own definition of a finished product. Extra Processing can also occur when tools and machines are not properly maintained and therefore, operate at non-optimum levels. Tools that are improperly cared for—thrown to the ground and not routinely cleaned or maintained—make the technicians' job more difficult. This prolongs lead times and adds to inferior quality repairs.

Transportation

Transportation is the act of moving parts, products and materials around the shop. The travel distance of parts, products and material can be vast, in terms of both distance and time. For example, the movement of parts and supplies from the storage room to the work areas or the movement of cars from parking lot to shop interior and back to the lot. Inadequate or poorly planned parking results in the continuous game of moving cars to unblock other cars. Poor or inadequate organization of storage facilities causes time to be lost when parts are moved to unblock other parts. Retrieval and travel time can be expansive, exhausting and costly.

When cars need to be moved, technicians must locate the keys, find the car, and then assess the situation. Many times, when the car is blocked, they must move other cars first, causing them to have to locate those keys and move those cars around, until they are able to finally reach the car they need to work on. Transportation of smaller items like parts and boxes may require a hand truck or pallet jack, which must be stored in an easily accessible location, and maintained in proper working condition. In both cases, transportation causes increased costs, which can be inflated by inadequate planning and inefficient processes.

Paperwork also travels and, therefore, is considered part of the cost of transportation. An estimate, for example, must be moved through the office and then, once completed, must be transported to the shop floor.

Motion

Motion is any movement of technicians without a productive result. Any time that a technician stops working in order to locate a needed item, a wasted action occurs. Searching for tools can be a major time waster, which inflates your costs, and which clients don't compensate you for. Poorly organized tools, or piles of tools, do not allow workers to quickly locate what they need. Walking around and sifting through piles is not productive. Also, sharing tools among technicians can cause a vicious cycle of borrowing, misplacing, losing and re-purchasing. Without a clear vision for shop organization, waste simply multiplies.

Other examples of motion in a shop are:

- Searching for parts and supplies
- Searching for office staff
- Reviewing information
- Walking to and from each area to solve a problem

Waiting

Waiting occurs whenever any process is out of balance and therefore, causes people and machines to stop and wait. Shop personnel can be required to wait for many reasons. Teardown technicians need to wait for the estimating process to be completed before they can begin work. Re-assembly technicians may need to wait while the car is in the paint booth. In a smaller shop, with fewer cross-trained workers, technicians with less experience may need to wait for experienced technicians to assist them or rectify problems.

The shop, as a whole, may need to wait for the parts room to respond to their requests for parts and, in turn, the parts room may have to wait on their suppliers (dealerships and part supply stores). Waiting is clearly a non value-added part of the repair process. The most appropriate perspective you can have regarding this is that the car (or customer) is waiting for someone to begin value-added work.

Defects/Rejects

Defects and rejects include any rework, mistakes, quality errors, returns or comebacks. Whenever there are humans involved in a process, mistakes will occur; it is unavoidable. However, the quantity and severity of the mistakes can be minimized with lean implementation. Waste breeds waste. Whenever technicians must stop performing value-added work and leave the area, in order to search for tools and information, they lose focus on what they are doing. A good dentist does not leave you sitting in the chair while he goes away to retrieve important dental equipment and devices. Everything he needs is at his fingertips, allowing him to focus on the job at hand, fixing your teeth. A technician's car repair environment should be similar, allowing him to focus on the required work, without variation or distraction.

Quality errors can also be made when standardized work procedures do not exist. For example, there is a preferred method of prepping and painting a bumper, which should be used by all technicians. Time and motion studies demonstrate that technicians will perform the same work, with varying levels of cycle time, quality and supply usage. Standardized work is the best practice of any given operation. To eliminate waste, it is critical that all work processes be standardized and therefore, repeatable by all workers.

Inventory

Excessive amounts of shop supplies, parts and material occupy much needed floor space, which results in lost money. In a manufacturing facility, inventory is often the biggest waste and is caused by excessive purchase of raw materials, unfinished product and an overabundance of completed units waiting for buyers.

There are three types of inventory; raw material or parts, work-in-process (WIP), and finished goods. Within an auto body shop, the following examples of wasted inventory exist: hardware, fluids, filler, and body panels (raw); partially sanded bumpers and unfinished repairs (WIP); and completed cars waiting on customers (finished goods). Finished goods in your industry can be a positive, since the cars are ready for pick up and customers love to have their cars sooner than anticipated. Yet, a completed body repair can be negative if another vehicle should have been finished first according to the delivery schedule.

Keeping raw material and parts to a minimum is very important, and this includes any supplies. While the lure of a discount for buying extra may be appealing, the cost of storage must also be factored in. Does it make sense to have three cases of Honda manual transmission fluid if your re-assembly process only uses one bottle every two weeks? Tying up money and space with excessive purchases is often compounded for dozens of items, which adds up quickly, affecting your bottom line.

Additionally, ordering parts for a car too far in advance will mean storing the parts while waiting for the car to arrive and the work to begin. Section 5 will cover The Lean Parts Room and explain a leaner approach to material replenishment.

Waste removal is the cornerstone of any lean journey. It is a continuous process of analysis and solutions for each individual process in the company. In short, it never ends.

Waste is evaluated based on its impact on the process. Some waste is difficult, or nearly impossible to remove, and perhaps, should be left alone. Mistakes in the approach to waste removal are often made by improvement teams and therefore, efforts should always be directed toward high impact waste. Waste should be removed based on priority, and also on how its removal will affect the overall process.

Non-Utilized Talents

Non-Utilized Talents is the act of not properly utilizing technicians to the best of their abilities. People are only as successful as the process they are given to work in. If a process inherently has motion, transportation, extra processing, overproduction, periods of waiting, and defect creation, then that is exactly what they will do. That is Non-Utilized Talents. Body shops that do not train and mentor their employees to become cross trained and experienced in other elements of the repair process are committing the 8th waste. Although it may appear that having someone working on the same tasks over and over will make them more experienced, you as an owner are not fully utilizing their talents. That is Non-Utilized Talents.

High Priority Waste

This type of waste should be identified and removed from the process as quickly as possible. Professionally speaking, I would classify the action of a technician leaving the work area, as high priority waste, in both action and associated time. By definition, technicians are considered value-added because they repair automobiles, which results in profitability for the company. Therefore, their time must be focused and directed toward the work taking place in their areas. Walking around, searching for tools, parts, supervisors, documentation, standards, supplies, or any other item, is high priority waste, and has an immediate, negative impact on the shop floor, creating bottlenecks, as well as animosity between technicians. Additionally, technicians who need to leave their work areas may lose focus and could, possibly, commit quality errors. To illustrate this point, let's take another look at the way a dentist operates. Each dentist has an assistant, who hands him/her the tools and attends to the patient's physical comforts. This allows the dentist to focus solely on the patient's teeth, and the important work he is performing. All the necessary tools and supplies are placed on a nearby tray, within easy reach of both the dentist and the assistant. The dentist and the assistant have no need to leave the patient to retrieve tools or other necessary items. In fact, leaving the patient during an important procedure would not be optimal performance, or adequate patient care. Similarly, when technicians leave their workstations, they are practicing high priority waste, which can have negative effects on the process, the product, and the company.

Medium Priority Waste

Medium priority waste is generally associated with waste that occurs while the technicians are within their work area. Technicians are often required to remove parts from the original packaging that came from the suppliers. Removal of foam, shrink wrap, tape, bags and boxes can consume valuable repair time, and also requires that garbage cans or recycling bins exist in the repair area. In regard to the eight wastes, this would be considered "Extra Processing." Removing packing material should be performed in the parts room, unless the parts need to be protected from damage.

Lack of timely information or insufficient communication can also cause unnecessary wait times. Technicians must have appropriate methods and signals to communicate their need when they are trying to locate an office staff member. Without adequate ways to communicate, they may opt to leave their work area, which would place them in the high priority waste category. Technicians, who are not able to communicate their needs, may continue to work without answers to their questions or necessary supplies, and this type of waste will continue until the appropriate person happens to stop by their work area. In some cases, technicians may even need to stop working and simply wait until someone comes by to address their need.

Low Priority Waste

The movement of the cars in the repair processes is non value-added, and an example of wasted motion. However, this movement is nearly impossible to remove from the repair process. Some manufacturing companies have implemented continuously moving assembly lines, where the operator walks as they install the parts. I do not recommend this, and believe it is not conducive to good quality and safety. To illustrate this, I will provide an example of what occurs when a line moves, although I am not advocating this type of process, but rather the separation of repair requirements by work area.

A manufacturer of vinyl windows attempted a moving line in one of their high-volume plants, where they operated approximately twenty-two manual and automated assembly lines. Some lines had conveyor belts, which were operated by the worker at the start of the process. He would finish his work on a window, place the window on the conveyor, and press the foot pedal to move the belt. The product moved and operators on the line were forced to move with it while performing assembly, regardless of where they were in their own assembly process. This reminded me of an old *I Love Lucy* show, where Ethel and Lucy were putting chocolate candies into boxes and couldn't keep up with the pace! An analysis of internal and external quality data at the window company revealed a significant amount of missing parts in the windows that flowed on the conveyor belts. Even more telling, the parts that were most often missing were those that had to be installed. Since installing required time, the operators simply skipped that step; no time to do it. I also observed many operators lying on the moving conveyor belt, attempting to install parts without regard for personal safety.

This story clearly illustrates that some waste is nearly impossible to eliminate 100% and that attempting to do so is neither expected, nor desired. Your focus should be on reducing high and medium priority waste, which will be a continuous process and effort, but will result in dramatic improvements to your body repair process.

I am certain that the wastes I have discussed seem familiar to you. The question is how much waste do you have?

Once a client has gone through their first training session on waste, I always ask them the following questions: "So, what do you think? How much time is spent looking for tools and supplies? How much time is spent walking around and retrieving parts and cars and moving them from one process to the next? Do you overproduce and buy excessive supplies and parts? How many mistakes occur in a given week? 20%? 30%?

For the purpose of this guide, let's use 30%, which, by the way, is a conservative measurement.

Available Time

Now that you have a concept about how much waste exists, begin thinking about how this waste affects time. *Available time* is the time people actually have to do their job. On the shop floor, it is the value-added time of actually repairing cars. Machines running and cars being painted is considered value-added time. However, delivering parts and supplies is non value-added activity; when parts and people are moving, nothing is being repaired.

Available time is slightly different from one shop to the next, but there is some commonality. To determine available time, you must subtract the time spent in breaks, morning meetings, lunches, etc. Here is an example of a traditional 8-hour day:

480 Minutes	In the Shop
- 10 Minutes	Morning Start Up
- 10 Minutes	Morning Break
- 30 Minutes	Lunch
- 10 Minutes	Afternoon Break
- 15 Minutes	Clean Up
405 Minutes	Available Time
OR	

6 Hours and 45 Minutes Remaining Work Time

Estimating performance, based on 8 hours, is the wrong approach. Following lean philosophy, if 30% of the day is spent in waste, your true available time to perform work or conduct value-added work is only **4 hours and 45 minutes**. This also means that 30% of wages and salaries are spent on waste.

Lean auto body techniques are very powerful in reducing waste and then, maximizing the time spent on performing the work your customer's want you to do.

Time is something we all cherish. While I am not implying that all 30% of wasted time could be recovered, as that would be unrealistic, a large portion of it could be saved. Lean is the way to make that a reality.

Section Two – Lean as a Business Model

The primary focus of any business should be its customers and how the customers' needs can be met. In the auto body repair business, there are, essentially, two customers: the car owner and their insurance company. All potential and existing customers have certain expectations and if body shops wish to satisfy and exceed those expectations and remain competitive, they must establish solid best practices. In today's market, people can go just about anywhere to get their cars fixed, which is why customer service has become such an important factor of doing business. The concept of lean manufacturing is no longer new; it has made its way through a variety of different industries and has been proven successful. Even so, most body shops have not fully embraced the lean philosophy or applied what they have learned about it to their daily operations. And yet, these same businesses may wonder why their shop performance is in jeopardy, as compared to their competitors. Athletic teams and athletes operate under the principle of continuous improvement, constantly seeking to find an advantage or competitive edge that will make them more successful as a team. A successful business operation is no different. In today's economy, three main business decision drivers exist, which can make or break a deal. Customers who seek a particular product or service will base their decision on cost, quality and delivery. The potential client uses these three metrics to decide if your product or service is worth purchasing. It is important to find an efficient balance between these drivers, focusing on all three equally, so that one is not given all the attention while another is allowed to break down. However, optimum balance of focus can be difficult, and each company defines it differently. The only way to ensure a focused balance is to employ the methodology of continuous improvement, which is the fundamental principle of lean auto body.

Lean auto body principles, when correctly applied to your environment, can be a powerful tool to determine optimal cost, quality and delivery for your products and/or services. Although some people may tend to look for the cheapest method, the ideal approach is to find balance, which takes focus and dedication. In the following sections, I will break down each of the three drivers, noting how they can affect each other within a body shop environment.

Cost

Typically, companies handle cost in one of two ways: cost cutting or cost managing. Those that favor cost cutting employ downsizing, firing, improvising, abuse of suppliers and cutting corners, all of which are clear cost cutting actions. Some companies even believe that lean manufacturing is cost cutting, which is simply untrue. Rather, lean is all about managing costs, which is much better business sense, in the long run.

Focusing only on cost, the quality and potential delivery of a product or service will suffer. In an attempt to lower cost of labor, many shops try to eliminate the number of hourly workers and run "thin" repair lines. Both of these measures constitute "cutting costs", but they are usually futile efforts. When the labor force is drastically reduced, the already scheduled technicians typically must pick up the slack and perform faster, in order to meet critical pick up dates. Essentially, the process is working beyond capacity and this is dangerous, as critical repair and quality steps may be skipped in order to meet demand. What usually occurs is that the delivery date cannot be met, due to a shortage of personnel, and the car is either completed late or with sub-standard quality, upsetting the customer and tarnishing the company's image.

As a lean practitioner and trainer, I advocate making process improvements without a lot of cost; however, there are certain necessary items that must be on hand to ensure an effective process.

Repair processes all require tools, workbenches, lighting, part presentation, documentation, shelving, bins, tool holders, equipment, fixtures, and jigs, and these items need to be working properly to allow the automobile to be repaired smoothly and efficiently.

The costs associated with training and mentoring are sometimes considered too expensive. This is not smart thinking. In contrast, I recall one circumstance in which I was negotiating with a company called Offutt Collision Repair in Bellevue, NE. While discussing our potential agreement, the owner stated that the cost of initial training was something he could not afford NOT to do. As far as he was concerned, training was critical to his success in an ever changing market. Needless to say, this company is currently prospering in an industry where those who have not embraced change are struggling. The owner is beating the competition and meeting the ever changing needs of his customers. Manage your costs, don't cut them.

Quality

Without quality, you have nothing. If I was forced to choose just one driver to keep at an optimal state, I would choose quality, hands down. Now, by no means should cost or delivery be allowed to fall apart because of quality, but, in my personal opinion, customers are more loyal to quality than to any other driver.

Like all of you reading this guide, we are consumers seeking products and services. With the occasional exception, of course, I truly believe that people are loyal to quality, above all else. If the quality of a product or service meets or exceeds expectation, we will sacrifice a bit on the cost or delivery. After all, it may be worth a little delay or additional cost, knowing that we can rely on that particular business to provide good quality products and services. I do realize that many consumers simply look for the rock bottom price, yet I still find that most value quality over anything else. Let me give you a personal example.

Like most others in America, I own a car, which carries a variety of associated costs such as gasoline, insurance and maintenance. As a consumer and automobile owner, I can take my car anywhere for repairs. There are countless options and each automotive repair company offers some level of cost, quality and delivery. That being stated, sometimes, we opt to have our buddy repair our car, rather than take it to a stranger; although most of us don't have that luxury. There is a quick lube shop down the street from my house, probably five minutes away, at the most. This shop offers a multitude of services and products and at a very inexpensive price (cost). They are very fast and can have my automobile done in a timely manner (delivery). The problem I have with this particular shop, which is not representative of other quick lube shops, is that their quality is not to my satisfaction. They also make a practice of hiring high school kids with limited knowledge of automotive repair. In my opinion, the quality of service is less than adequate. Therefore, although this shop appears to balance the other two drivers quite well, quality suffers; and I am loyal to quality. So, what is my alternative?

I choose to take my car to a repair shop farther away from my home. This shop tends to take a longer time to get the work done (delivery); I often have to wait in the customer lounge, reading magazines. Their prices are higher than most shops in town (cost), but their quality is outstanding. They have fantastic customer service and I never have an issue in regard to their quality of work or documentation. Basically, I am willing to put up with a longer delivery time and higher cost, because I know the quality of their service is worthwhile. However, if cost and delivery got farther out of balance, I might be tempted to seek another vendor. That is a real possibility.

Now that you have a good image of quality, I would like to discuss what happens when too much effort is spent on quality of product or service. In my repair shop analogy, I was happy with the balance of the three drivers, although this was a personal preference, and may not reflect how other consumers might feel. But, whenever there is excess focus on quality, cost and delivery can suffer.

Body shops that focus most of their attention on quality may, in fact, not trust their ability to repair a car. Perhaps there has been a history of poor quality and they are afraid of losing customers. Companies that spend a lot of time on quality have a great chance of increasing their operating costs. Assigning more people to focus on quality is expensive, and at some point, it becomes redundant. Inspections and testing are not considered value-add activities, unless the customer is willing to absorb the cost. Some manufacturers have government contracts that demand strict adherence to specific criteria, but the government is willing to pay extra for that. In this particular case, it may be cost effective to perform repetitive inspections and testing, but most manufacturers don't fall into this category. Delivery is also negatively affected by an over-emphasis on quality checks, causing extensive lead times due to the extra time needed for checking and re-checking product or, in your case, an automobile.

Delivery

Are you volume driven? Fix, fix, fix seems to be the mantra of many body shops. In my twelve years as a lean practitioner, I have not yet come across a manufacturer or auto body shop that was not output driven. This is not stated to imply that making deadlines and delivering the product on time are bad things. It's the ridiculous notion of speed that seems to be the cause for much of the chaos. Body shops that focus on delivery, exclusively, will very quickly see an increase in cost and a decrease in quality, especially in uncontrolled working environments. Companies that have this volume driven mind set, cannot see any further than the next hour; there is no long-term vision for the day, week or month, and crisis management is the name of this game. The first indicator of an inefficient repair process is a shop supervisor or manager who "throws people at the process" to make daily repair requirements. Adding technicians to the process, without reason, costs money and will not add any inherent value to the repair flow of the car. The cost of labor simply increases, as does the risk of potential quality problems such as scrap, rework, and future customer complaints. It is hard for many managers and technicians to break this habit, even after lean principles have been applied, but it is a mentality that should be reversed, for many reasons. First, the added cost is not usually passed onto that first customer, but is typically added into the overall business costs, which will affect future customers. Second, it is a cost that the shop must absorb initially and therefore, funds may be pulled from other areas for coverage, causing shortage in the areas of deduction. Business 101 is to establish your organization to accommodate the original price agreed upon with the car owner and insurance agent, while being able to fund operating costs. Therefore, you are shooting yourself in the foot by creating added costs outside of your business structure. Additionally, you will need to prepare for the external costs that could arise through customer complaints, returns, and service calls.

Simply put, there is no gain to an organization if we simply create problems and add cost by driving processes and technicians to meet unrealistic pick up dates. Repair processes should be designed to meet delivery dates effortlessly, and should be monitored continuously to meet that objective.

Balance is the key with cost, delivery, and quality. Achieving optimal balance in these drivers will be an ongoing challenge. Lean auto body provides a variety of tools and principles that can be applied to a business to help achieve optimal cost, quality, and delivery. Your first step in getting there is to recognize that the customer can go anywhere to do business and, in fact, may already have done so. Customers are the reason we exist, so remember that they have a choice. And, their choice will always be based on cost, quality, and delivery.

Section Three – Lean Auto Body Metrics

Commitment to change is essentially a philosophy, but it is the key to a successful lean implementation. Change is not easy for some individuals. In fact, for some turns the familiar into the unfamiliar and therefore rocks the boat. Getting people to embrace change as a positive measure can be very challenging and there is no perfect formula to follow. Change is extremely difficult, as it often involves people adapting to change or teaching others to adapt to change. The challenge lies in getting everyone to see change as valuable and necessary and ultimately, the best insurance of success. Effectively, change is not in our words, but in our actions.

To avoid lean implementation failures on the shop floor, management must first establish the foundation for success, by developing goals and metrics to improve **cost, quality, and delivery**. Improvements in these areas will have a profound impact on the company's financial strength, as well as overall growth. Actions should always follow words. Don't simply state, "We are doing lean"; demonstrate it, by taking action. This important concept is what I call the strategic purpose.

The strategic purpose is an effective way for management to demonstrate their commitment to the lean program and to the lean philosophy of continuous change and continuous improvement; beyond just inspirational speeches and company declarations. The strategic purpose serves as a guideline to implement lean processes and sustain positive change. Some individuals refer to the strategic purpose as a lean strategy.

In order to help the business overall, every company needs to establish a list of critical shop floor metrics that can be measured and quantified. These metrics are often called KPIs, or key performance indicators, and are listed below:

- Productivity
- Quality
- Inventory/Work-in-Process (WIP)
- Floor Space Use
- Throughput Time
- Productivity

Productivity

Productivity can be measured in a variety of ways. Productivity is improved when automobiles are repaired with less effort, less manpower, less equipment, and less utilities (overhead), for example. Since lean is about managing costs, not cutting costs, body shops need a smart approach when attempting to achieve minimum effort. Supervisors and managers always seem to be concerned about speed, but productivity is not about speed, it's about pace. Over the years, manufacturing and auto collision repair professionals have had a misconception about productivity. Working hard, fast, and to excess, is not conducive to good quality and safety. Technicians can only sustain 100% speed to a point without negatively affecting quality or seriously negating safety factors. Lean is about working smart, at a pace that is sustainable, while safely completing the required number of good quality repairs in a given time period.

Productivity is directly related to cost. Therefore, it should be the most important shop floor metric to have in your strategic purpose. Focusing on the efficiency of the technicians is critical because they are considered value-added; they build product that, in turn, pays the company's bills. With a more efficient process and less personnel, cost will be dramatically improved. Less personnel equates to less labor, less tools, less workstations, less documentation, and less material, as examples. Fewer people handling cars equates to improved quality. These are key points that contradict the old philosophy of simply throwing more bodies onto a line in order to meet deadlines. Unnecessary people working on the product, with an uncontrolled and poorly designed process, can increase the chance of quality errors. Not to mention, safety issues!

Working in a lean environment means LESS! Less processes, less people (proper manpower allocation), fewer transactions and systems, fewer hands the car must travel through before landing in the customer's hands. Delivery will be improved substantially. The key is to continue to manage costs and re-deploy personnel, as needed. As areas become more proficient, due to improved processes, implement appropriate modifications. Don't simply cut people—that is not the lean philosophy. Proper manpower allocation is always the best approach.

Quality

Quality is one of the three main drivers, and I recommend that it also be one of the key shop floor metrics in your strategic purpose. However, before adding quality to the list of metrics, you must decide how quality will be measured. Many shops, both large and small, do not measure quality and their profits have suffered because of it. Quality should be measured both internally and externally. Be sure to track the details about returns as well, as sometimes a customer returns due to unsatisfactory work, or with a dent or scratch received after their initial repair work was completed.

The most important quality measurement is your internal metrics. Track any hours that fall outside of the original estimate. Internal metrics are better indications of quality and can be used as a baseline for improvement.

Inventory/WIP

A lot of money is invested in parts, supplies, and material, which is why they need to be included in your metrics. Accurate metrics will provide a method for monitoring waste and making improvements. However, reduction of inventory doesn't happen overnight. Whenever I tour auto body shops, I always observe fluids, filler, adhesives, rags, paper towels, zip ties, hardware, sealant, and various other shop necessities, piled all over the shop room floor, taking up valuable work space. In addition to this, I find parts rooms completely overstocked.

Parts room supervisors may be enticed by vendors to make overly large purchases in order to get discounts. For example, purchasing seven cases of windshield cleaner, that will take several months to deplete, just to get a 20% discount. While the discount was attractive, the surplus cleaner then needs to be stored, and if the parts room is already filled to capacity, the logical storage location is the shop floor, where it takes up valuable space and is in danger of breakage and leaking. Excessive purchasing of unneeded inventory drastically affects your profitability and should definitely be eliminated.

Floor Space Use

Floor space comes at a premium. One of your highest overhead costs is the rental, leasing, or purchase of your building. Utilizing floor space efficiently is critical to the success of any lean journey, especially in regard to cost, quality, and delivery. The shop floor serves one major purpose; to repair damaged cars. Although the shop is used for other purposes, such as holding inventory and supplies, detailing and maintenance, the shop floor should primarily be used for value-added work; car repairs. Non value-added work on the floor means less profit for your company, especially considering the high cost of owning, renting, or leasing the building. Over time, shop floors can become cluttered with an accumulation of workbenches, refuse cans, chairs, machines, tables, carts, tools, parts, and pallets—all of which take up valuable real estate. Therefore, including floor space in your strategic purpose is critical. As metrics are monitored and improvements occur, valuable space will become free, allowing body shops to add more repair lines, repair more cars, and become more profitable.

If vacant space exists, without clear, visual objectives for use, the space will quickly become filled. That is just human nature. Unnecessary items will begin to pile up and consume valuable repair space on the floor or in the work areas.

Cost is significantly improved when floor space is better utilized. Adding repair lines to accommodate more cars will increase revenue, as long as the cars aren't sitting idle for extended periods of time. If metrics determine that less floor space is required, the shop could move to a smaller location, saving money on rent or finding a more advantageous lease agreement. Why pay thousands of dollars for rent if your shop floor is only 50% utilized? Quality is significantly improved when less space is used to store unnecessary items; items that could become damaged or broken. Also, less clutter in the area promotes better visibility throughout the shop. Floor space is expensive, use it wisely.

Throughput Time

Sometimes used in conjunction with travel distance, throughput time is the time it takes the car to go through the entire repair process. Obviously, throughput time has a direct impact on pick up dates—the longer it takes to repair the car, the longer a customer must wait.

Improving key shop floor metrics will have a profound impact on the overall financial success and long-term growth of the organization. The metrics discussed earlier are particularly important because they are directly related to the work being performed on the shop floor; work which affects profit and ultimately, employee paychecks. Technicians need to work in an efficient environment in order to be successful contributors to optimal cost, quality, and delivery.

The shop floor is only one part of the overall lean strategy. With a strategic purpose in place and metrics established for the shop floor, an organization has a good feel for why it is in business. While each department should eventually develop their own internal metrics for improvement, when it comes to developing an all encompassing strategic purpose, the shop floor is the key.

Section Four – 5S and the Visual Shop

Every company that decides to embark on a lean journey faces a common struggle—where to begin. Body shops are no exception. Not all lean tools are applicable or beneficial to every industry, which is why a lean specialist is typically there to assist with strategic implementation. There is no perfect roadmap or lean textbook that can be used as reference. First, you must become familiar with the tools and then, decide how they apply to your company and/or organization. Specific lean tools, when applied to the auto collision repair industry, will dramatically improve business metrics. 5S is one of them.

5S is an aggressive organization and cleanliness philosophy that creates order and discipline, resulting in a beautifully organized visual shop. Imagine attempting to drive your car across the country with no visual aides to direct you. It would be an impossible feat. Making your shop into a visual shop is the first step in becoming a lean practitioner. The visibility of your shop will prove to be an extremely powerful factor in enhancing performance.

The benefits of 5S include:

- Reduced shop cycle time
- Safer environment
- Available floor space
- Reduced waste of motion and overproduction
- Creation of a visually pleasing shop
- Visibility of potential issues

Now that you are aware of the benefits that can be realized from 5S, I will describe the importance of the 5S concept and the visual shop. The 5 S's are:

- Sort
- Set in Order
- Shine
- Standardize
- Sustain

Sort

Sort is the act of removing and discarding all unnecessary items from the shop. The sort function also identifies items in the immediate work area that may be necessary at some point in the process, but should be located or stored elsewhere. Sorting can be difficult for many people. It is a natural tendency for many of us humans to collect and store stuff, even stuff we truly don't need! That's why the self-storage industry is a booming business. A lot of time and money is invested in the storage of our unused and unneeded possessions—things we just can't bear to part with. That same mentality can exist in any business, including auto collision repair.

One of the most effective ways to perform the sorting function is through a red tag event; an organized approach to sorting, which allows a team to make group decisions about what items need to be eliminated from the work area. A red tag event includes three things:

- Red Tags
- Red Tag Area
- Removal Procedure

Red Tags

Refer to Image 4.1. Red tags contain a variety of information describing the item in question. Red tags are completed and placed on the items deemed unnecessary. The tag provides a visual indicator, identifying an item that will be removed. Start working in one specific area, identifying and tagging items for removal. It is also a good idea for the team to develop an inventory list of the items to be removed, for tracking purposes and to allow for future decisions regarding discarding or storing in other areas. The decision making will occur in the third phase of the sorting process, which is known as removal.

Figure 4.1 Red Tags
(Product image courtesy of Enna, www.Enna.com, 360-306-5369)

Once the team has effectively sorted the entire area, it will be filled with red tagged items. Remember, in this phase, no decisions have been made regarding what to do with the tagged items; they are simply marked as unnecessary for the respective area. By the way, if you find items destined for the garbage, toss them, don't tag them!

Red Tag Area

Once the team has completed the red tagging, remove the items and store them, temporarily, in a specified area of your shop. The temporary staging area allows the company to evaluate the collection and make decisions regarding the fate of each item.

Removal Procedure

Do not place your shop into the cycle of holding on to junk or moving junk around. Every shop has dumping grounds, where things accumulate and pile up over time, only to be moved around when the piles become too large. Instead of getting caught in this trap, develop a strategic removal procedure for your shop, which identifies appropriate options for removal such as the following:

- Garage Sale
- Recycle
- Donations to Schools or Other Shops
- Saved for Future

Although discarding is an option, don't simply toss away unneeded items without thinking about selling, donating, or recycling. Some items may even be suitable for your own future needs. However, the bottom line is that you only want to have essential items in your shop. Body shops have complex functions and so, there may be items that must be stored, even though they are rarely used. During the sort phase, simply identify those items as items that need another storage location. Remember to organize your work areas based on the rule and not the exception. This will efficiently allocate your shop space and allow for maximum efficiency and low cost, which equates to profitability.

Set-in-Order

The second phase of 5S requires more time than the first phase. This is the step in the process where the tangible and visual elements begin. You will need the following supplies to get started:

- Pegboard
- Peg Anchors
- Paint Pens
- Colored Floor Tape
- Spray Paint
- Label Maker
- Laminator and Sheets
- Stencils

These items are used to organize and visualize the shop. You will add other items to your 5S supply list, but this is a good start for the purposes of this guide. I also recommend that you either purchase or build a 5S supply box, to keep your supplies organized and available for all 5S projects. The set-in-order phase takes the longest and can often be the most tedious. The goal of this phase is to organize the area, assigning a home for each item and thereby, improving the flow of incoming parts, supplies, and material. Of utmost importance is the need to reduce waste of motion and the transportation associated with searching for needed items.

Always enter this second "S" with the purpose of making everything in your shop visual and accessible. Opening floor space and work surfaces is the key to any 5S implementation. I always advise my clients to "go vertical" and use dead space effectively. Dead space can be found throughout your shop and it is important to use it wisely. Flat surfaces are not the only locations to store and hold items or supplies. Before you add more workbenches and tables, take a good look at the space on your walls and in the corners. Almost anything can be placed vertically on tool and supply boards, just by using a little creativity. Avoid using cabinets, drawers, and shelving as much as possible, and go for open, hanging options instead. Peg boards are ideal for this type of storage. Save your drawers and cabinets for delicate items such as calibration devices and hang the more durable items where they can be easily viewed and retrieved.

Start with your floor items first. Decide which surfaces in the work areas must be kept open for actual work to be performed. Evaluate the necessary size, based on the job that is to be performed, and begin placing these items, or whatever might be needed, into the work space. Typical floor items include:

- Work surfaces: workbenches, tables, etc.
- Garbage cans: identify the minimum needed, taking into account point of use (in order to reduce motion)
- Locations for staging incoming and outgoing parts
- Misc. items: vacuums, welding devices, specialty equipment/machines

After you position the above, observe the surrounding spaces and begin thinking about the most logical and practical places for the necessary tools and supplies. Here is where creativity comes into play. At this point, you are only drafting your design and nothing is permanent, so don't make any designations or identifications in the locations. You are still in the planning stage.

Any items that will be stored on the floor should be the first items placed in the work area. Then, lay out all the necessary tools, supplies, and other workstation items, in order to see what needs to be organized and stored. Think vertical and avoid flat surfaces, as you decide where to place each item. Challenge your team to come up with innovative approaches to organization. Try to work with the space as it is and only, as a last resort, opt to bring in another cabinet, shelf, or tool chest. The key to 5S is visibility. Tool chests and cabinets do not provide visibility and therefore, you cannot easily see where things are located or what items may be missing.

Break out the 5S supply box, peg board, paint, cleaning supplies and any miscellaneous items such as nails and hooks. Give your shop a thorough cleaning, followed by a fresh coat of paint throughout the area. A new coat of paint provides a fresh, clean appearance to the showroom, so try painting workbenches, shelves, and metal stands, in addition to the shop walls.

Once you have made firm decisions about which floor items will be placed on the floor and where they will be placed, it is time to mark and then "address" the floor, creating permanent designations for the items that will hold supplies, tools, parts, and other materials. Using floor tape, make an outline around the moveable, non-stationary items on the floor, identifying their location. Items which are bolted to the floor and heavy workbenches do not need to be outlined. If the item is in a greasy area, you can paint an outline, instead of using tape. Once all outlines are completed, you will create the designations/addresses. Figure 4.2 shows an overhead view of a work area with floor identification.

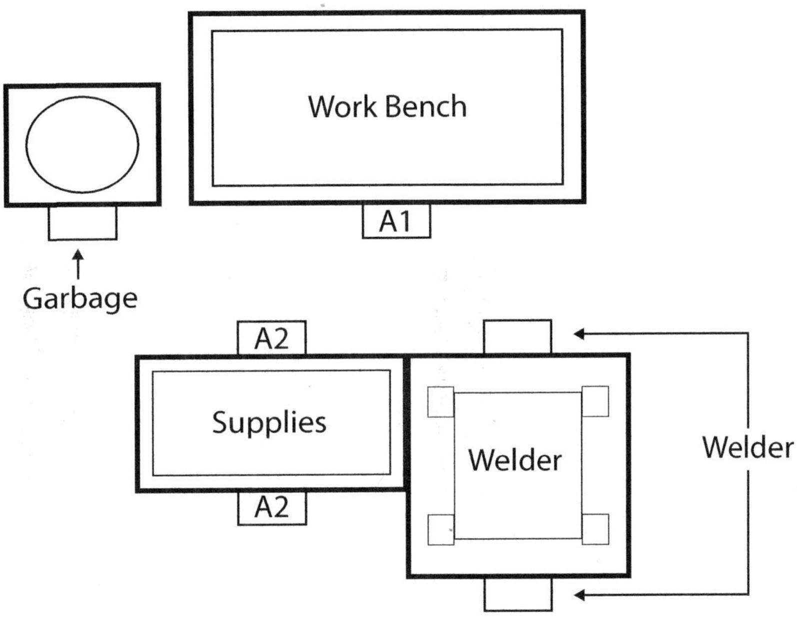

Figure 4.2 Floor Identification

As you can see, there are clear outlines of all items on the shop floor. The team should place labels or stencils on the floor, describing the address. In this example, A1 and A2 are being used. A1 and A2 are locations for important work-related items that are needed in the workstation. By creating this addressing, accountability is developed in the workplace. Depending on where the item is placed, like the welder, two labels/stencils may be needed. So, if a calculator, tool, or bin full of material goes in A1, it is physically marked "A1" on the item.

All items such as garbage cans, jacks, dollies, welders, etc., must have a home location and be clearly marked and labeled. Multiple garbage containers can have designations such as G1, G2, G3, etc. As long as you approach the set-in-order phase with the mentality of "everything has a home", you will accomplish the goal.

Addressing is critical. If, for some reason, an item is removed from its work area, it can be easily returned to its appropriate location. "Addressing" is an extremely powerful method of saving time and reducing motion and confusion. It can also be instrumental in helping to reduce the cost of repurchasing and replacing missing tools and supplies. Place your address designation labels high enough to be seen from a distance, to make them visible across the shop floor.

Tool Boards

Tools should be placed vertically on tool boards, where they are clearly visible and easily accessible. Often called shadow boards, these boards allow the work surfaces to remain clear and give the technicians an easy view. You can get very creative with these tool boards. Here are some tips for you:

- Paint the board: Any color will work; red, yellow, black, blue, green, orange, etc.

- Tool layout: Once the boards are dry, place them on the floor or another flat surface and lay all the appropriate tools and supplies on the board. These boards are excellent for holding items such as tools, scissors, calculators, clip boards, and other shop necessities. Your supply box should contain pegs, double-back tape, Velcro tape, and other items designed to hold things vertically. Take those things into consideration as you make decisions about where items should be placed on the board. At this point, you are designing your board layout; identifying how much board surface will be needed to hold the workstation's supplies.

- Cut the peg board: Now, remove the tools from the peg board and cut a square piece that is large enough to accommodate the tools for the workstation and will fit on your allocated wall space.

- Install the tool board: Construct a wood frame for your board, using 1 x 1 studs. The peg board needs to float at least one inch from the wall, in order for the pegs to have clearance and lock into place. Walls are the logical choice for tool boards, but actually, they can be placed anywhere. I have seen them installed on the side of workbenches and cabinets, and have also seem them on casters (wheels). Just be sure that your board is visible, accessible at point of use, and does not impede or disrupt the technician.

- Hang the items: Have your team hang the tools and supplies on the tool board. Be sure to leave room for labels in between the items and space to draw an outline or shadow of each item. This process is relatively slow, but it is the key to any 5S implementation. After placing the tools on the board, use the paint pens from your event supply box to trace an outline of each tool on the board. The painted outline is very visible when the tool has been removed for use or is missing. Next, make a name label for each item and place the label near the item's location on the board.

- Name/address the tool boards: Give each tool board a designation such as B5, J7, L3, etc. Label the tools with this board designation, so everyone in the shop will know where the tool should be located.

Images 4.3 and 4.4 show an example of a tool board that was built using these six simple guidelines.

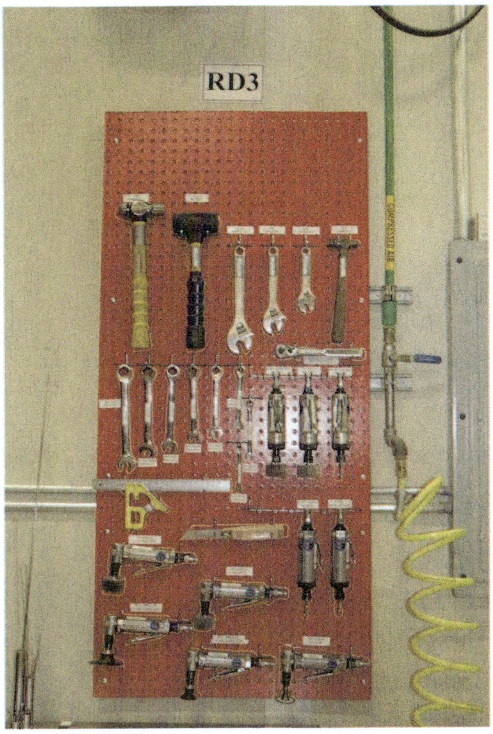

Figure 4.3 Tool Board

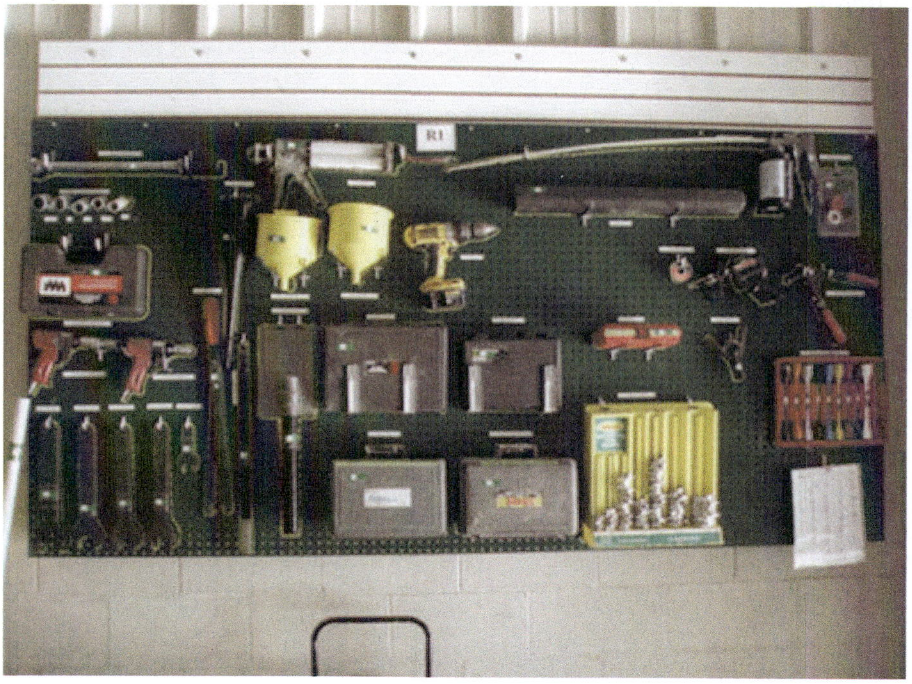

FIGURE 4.4 TOOL BOARD

The challenge with tool organization in the auto body repair industry is that most technicians are required to supply their own tools. Basically, the tools are their personal property and if the technician quits, their tools go with them. While tool boards make sense for company-owned, community-use tools, what can be done with personal tools? This issue is always raised during 5S implementations in body shops. One common rule of thumb is that a company cannot touch a technician's personal tools. However, there are some options and some important questions to consider, as the shop owner. Although the tools are not yours:

Who owns the floor space on which the tool chest is located? The company does.

Who owns the time spent sifting through piles of tools in personal tool chests? The company does.

Your business time and money (labor hours) is tied up in the technicians' personal tools. Disorganization and improper use of shop space causes waste, which equates to lost profit, which equates to loss of jobs. When technicians begin to see the value in visual boards, reduced clutter, and more efficiency in locating needed or missing tools, they will embrace the lean philosophy. To reduce excess, they will begin to take their unnecessary tools home, separate their tools into categories, and designate specific drawers in their tool chests. They will label their drawers and cabinets for easy identification of the contents. Some technicians will even take this concept further and develop "shadow drawers". Clear location and designation within a drawer allows the technician to have more organization and be able to easily see when a tool is missing. Start applying lean to your shop and encourage your technicians to take the lead with their own personal equipment.

Soon, the idea will catch on and your technicians will begin to take pride in their ability to organize and excel.

Parts, Material, and Supplies

The team must continue organizing all remaining items. Continue to think about reducing special needs. Use square bins to organize items that cannot be placed vertically on boards. When placing anything on a flat surface, be sure to mark off the area using smaller width colored tape. Mark all items with their home location and place all labels on the flat surface next to where the item will be located. *Everything* has a home...*Everything*.

Shine

I won't spend too much time on this one. During your 5S implementation, make sure to clean the area thoroughly upon completion of the set-in-order phase. Sweep and mop the area, wipe down work surfaces, and clean tools. It is usually a good idea to repaint any hand trucks and equipment, using the spray paint in your supply box. It simply provides a much more appealing and fresh-looking work environment.

Standardize

Be consistent in your approach. At the end of this section, you will find more 5S pictures that illustrate the consistency I speak of. All tool boards should look about the same, especially if you follow the step-by-step approach outlined earlier. Use color consistently; colors on the floor should have the same meanings, regardless of their location. Labels and labeling should be consistent. Think of it this way. Stop signs are red and shaped like a hexagon everywhere you drive in the US. Dashed lines indicate that passing is allowed and solid lines mean passing is prohibited. If we didn't have standardized visual traffic signs, we would not be able to travel from one state to another with ease.

Sustain

The final phase of 5S, after your first aggressive pass, is to create and enforce a sustaining program. Develop an end-of-day clean up procedure for each area, outlining the fundamental clean up and organization needed to complete the day. I also suggest developing some form of weekly auditing system, to score and evaluate the area after 5S implementation. Keep it simple and relevant, and be sure to tell your technicians that the purpose is continuous process improvement, rather than a method to point fingers.

5S can be one of the best lean tools you can implement, sustain, and continually improve upon. It is extremely powerful in reducing overall throughput for the cars; reducing cost, opening up floor space, creating extreme organization, and making your shop look very professional.

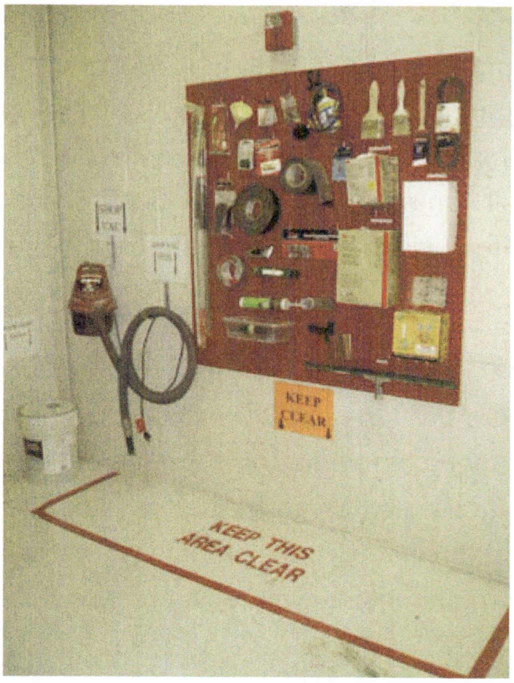

Figure 4.5 Parts Room

Figure 4.6 Floor Locations

Figure 4.7 Re-Assembly Area

Figure 4.8 Machine Shop

Section Five – The Lean Parts Room

If applied correctly, lean can have a profound impact on a company's inventory and supply levels. Some of our manufacturing clients have seen their inventory levels reduced by millions of dollars. Their factory supply levels of gloves, glues, rags, tape, and other items, have minimized dramatically. They have been able to turn newly accessible warehouse space into value-added production lines or fabrication departments. For our smaller clients, the positive impact was simply less money used for large quantities of inexhaustible materials and supplies.

The purpose of inventory reduction is not to reduce quantities to insufficient levels that will deplete too quickly and halt the repair process. Rather, it is about increasing the working capital, opening up floor space, reducing quantities to an appropriate level that allows for smart replenishment, and promoting enhanced visibility of potential inventory issues. This section will focus on your parts room. If your shop does not currently include a parts room, this section also provides an intelligent approach to establishing a parts room, from a lean perspective.

5S and Kanban

A lean parts room for the auto collision repair industry is the primary control mechanism for incoming materials, parts, and supplies. Its secondary purpose is to act as a material handler for the shop floor; similar to a material handler in an assembly plant, whose role is the delivery of parts to the workstations.

By now, you know the concept of 5S and the visual shop. 5S must be implemented into the parts room, just as it was implemented in the work areas. Another lean concept of utmost importance is *Kanban*; the Japanese word for "signal". Kanban is used in various ways, in every aspect of lean. In regard to inventory and material management, Kanban is a signal that more material must be ordered. In this section on the parts room, I will focus on these two lean techniques; 5S and Kanban.

5S in the Parts Room

The layout of the parts room is important to its functionality, as well as to aesthetics and organization. The layout should be determined based on the most appropriate flow and function. Here are the locations/areas needed:

- Incoming Parts Staging
- Dealer and Supplier Return Staging Area
- Staging for Outgoing Parts for Technicians
- Supply Area and Kanban System
- Tool Area

Incoming Parts Staging

Incoming parts need a staging area in the parts room. This can be a specific place on the floor or a specified shelf in the room. Incoming staging areas will vary in size, depending on the type of parts. For example, bumpers and hoods are large objects and will require a sizeable area. Keep the size and quantity of your parts in mind when identifying space for your staging area. Mark this area with floor tape and label accordingly. I recommend labeling on the floor as well as overhead, or on the walls.

The incoming parts staging area should be highly visible and seen by anyone entering the parts room. In addition to its primary function, this area also provides great visual management. A staging area that begins to fill to capacity is an indication that parts are not being processed, parts are being purchased/replenished too early or in excessive quantities, or that work issues may exist in the repair area.

Dealer and Supplier Return Area

Returning parts to dealers and suppliers is a common occurrence. Hopefully, the returns are not due to quality issues, which would be a much bigger problem. Since returning parts to suppliers is a function of the parts room, a small area or shelf should be reserved for this purpose. A visible, easily identifiable area for damaged or unneeded parts will reduce confusion among technicians, thereby reducing motion, transportation, and mistakes. The returns area allows the supplier to easily locate the parts that are being returned, without having to wait for a technician to assist them. Suppliers can quickly stop by your shop, check the returns area, and go about their business. The two most important factors are to designate a single area for returns and to identify it clearly with signs and labels.

Staging for Outgoing Parts for Technicians

The third function of the parts rooms is to prepare parts for the technicians and the cars that require them. Parts for a specific car should be placed on their own mobile rack and identified. The racks serve as temporary homes for the parts for that individual car, therefore parts for multiple cars should never be place on the same rack. It is also a good idea to develop a numbering system for each car that comes into your shop. The numbers can be used on the carts for identification. Figure 5.1 illustrates this point.

FIGURE 5.1 PARTS CART NUMBERING

Mobile carts are prepared by the parts supervisor, based on pick up dates and hierarchy of importance. The carts should remain in the parts room until the car is in the re-assembly area, ready to be worked on. Bumpers or other large items can hang on a hook on the back of the cart. You can purchase these hooks from many body repair supply companies. Since the carts are identified with the specific number associated with the car, the same visual management should appear in the re-assembly area; a one-to-one match. If you have the room, establish individual lanes with floor tape, label the floor with the number, and place a label or designation on the wall, as well. If only ten cars can be repaired in your shop at one time, establish only ten lanes. Figure 5.2 illustrates this point.

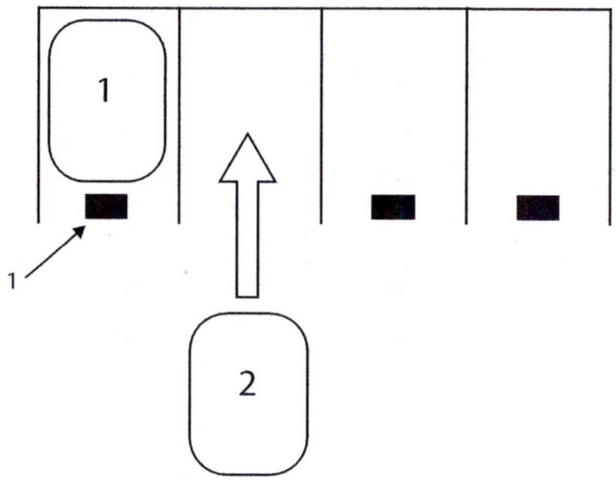

FIGURE 5.2 CART STAGING

Each cart is assigned to a lane, by number. When that cart is emptied, it is returned to the parts room, where it is filled with a new set of parts for a different car. To ensure that technicians are not wasting time retrieving these carts, you should develop a Kanban system to signal the parts room for more carts. I will discuss this system in Section 7, as well as signals for all shop floor materials and supplies.

Supply Area and Kanban System

When establishing a material a replenishment system, which includes all body shop supplies, set levels that will keep the quantity of materials and supplies to a minimum. Obviously, this applies only to items that are used on a regular basis, such as fluids, and not parts that are ordered based on car repair estimates. Items include:

- Vital Fluids (oil, coolant, transmission fluid, brake fluid, etc.)
- Rags
- Cleaners
- Adhesives and Glues
- Filler
- Sanding Discs
- etc.

The above are examples of materials and supplies that are used on a regular basis and must be kept in stock. The challenge is determining the correct amount to keep on hand. For example, how much Honda manual transmission fluid should you retain? How long will two cases last? How often does your shop repair Hondas, and how often do the cars require additional transmission fluid? I estimate that unless your shop only repairs Hondas, two cases will last quite a while.

The parts room is similar to the "stock" room in a manufacturer's warehouse. It contains vital supplies and necessary material to support the shop. Some supplies may be used for car repairs and some only for general use or facility maintenance. Regardless, all supplies must be set up on a Kanban system to monitor usage, keep cost down, and reduce carrying needs.

Here is a simple approach:

1. Identify all the material and supplies needed in the parts room
2. Separate them into categories (i.e., fluids, adhesives, sanding, general maintenance, electrical, hardware, etc.)
3. Identify the amount to be on hand at all times. Ask this question: How much is one week's worth?
4. Identify the re-order quantity: 1 bottle, 2 boxes, etc.
5. Decide where the supplies will be located in the parts room
6. Implement 5S for the supplies, so that each item has a home location, regardless of size
7. Place labels to designate the items.
8. Print out Kanban cards and place them near the item. (Refer to Figure 5.3).

Honda ATF Fluid

5678-98

Min: 1 Bottle

Reorder: 1 Bottle

Location: CB-1

Figure 5.3 Kanban card

Notice that the Kanban card includes the item description and part number, the minimum quantity on hand, the re-order quantity, and the item's home location in the parts room. When the minimum quantity is reached, that is the signal for replenishment. Sometimes, another visual signal, such as a red dot under the product, can be used to indicate that the card should be submitted for replenishment. When the "signal" is present, the card should be placed into a Kanban card bin. The card is now the signal to order the supply. You may want to wait until a collection of cards is accumulated, and then place the order. Images 5.4 through 5.9 illustrate this process.

Figure 5.4 5S and Kanban Cards

Figure 5.5 Kanban Cards have Numbered Sleeves

Figure 5.6 Replenishment Signal

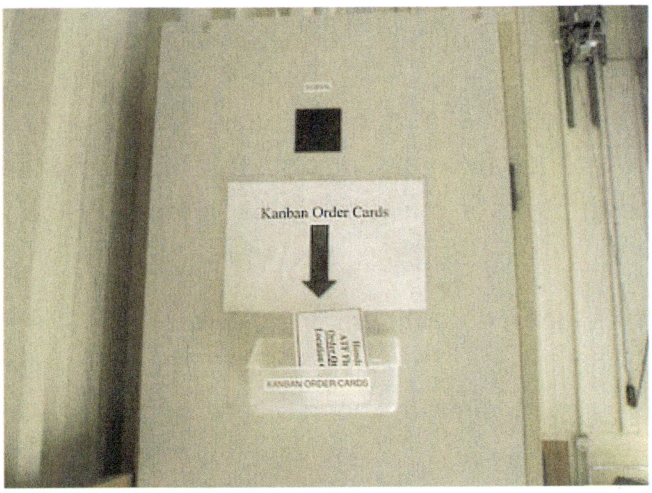

Figure 5.7 Place Kanban Card in Order Bin

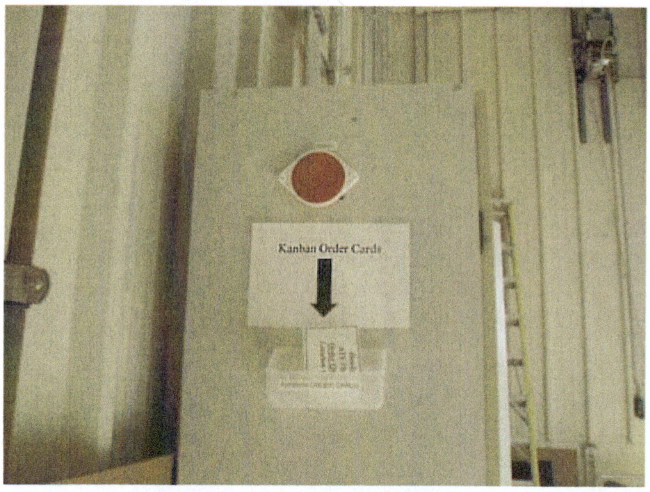

FIGURE 5.8 SIGNAL TO PARTS ROOM

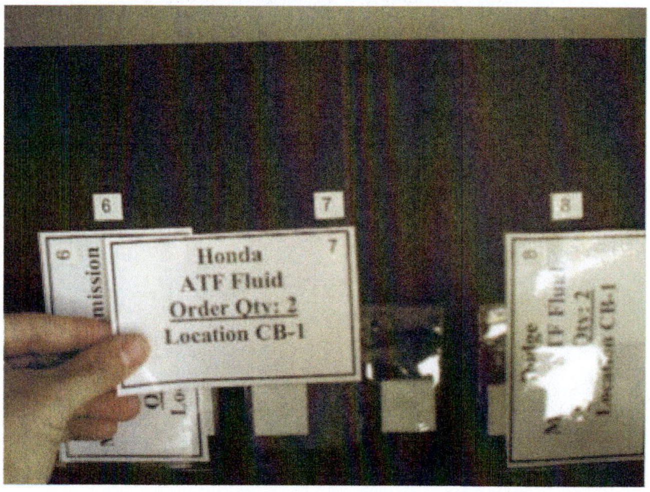

FIGURE 5.9 ITEM REPLENISHED AND KANBAN CARD RETURNED

Tool Area

Parts rooms generally store backup tools for the repair processes and miscellaneous tools for facilities work. Now that you have an understanding of how to organize tools with the 5S approach, apply the same techniques to your parts room. Follow the guidelines in Section 4.

The parts room is a vital support department. Inventory, supplies, material, and tools enter and exit the area on a regular basis. This must be controlled and therefore, the parts room must be correctly set up and monitored. I hope this section has given you some ideas and insight for that purpose.

Here are some pictures of a body shop parts room, as well as some other non-body shop Kanban examples.

Figure 5.10 Kanban Parts Room

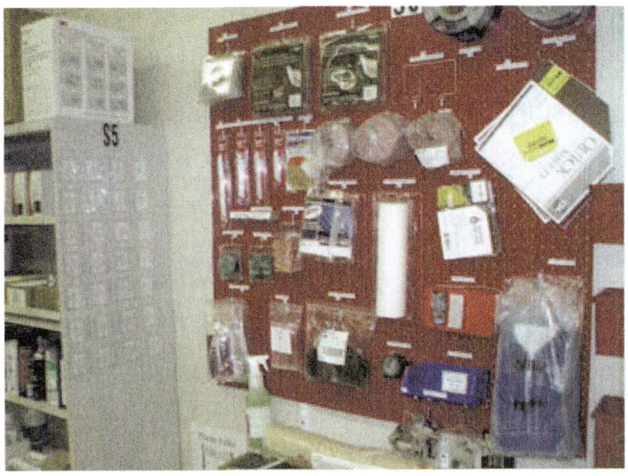

Figure 5.11 Kanban Parts Room

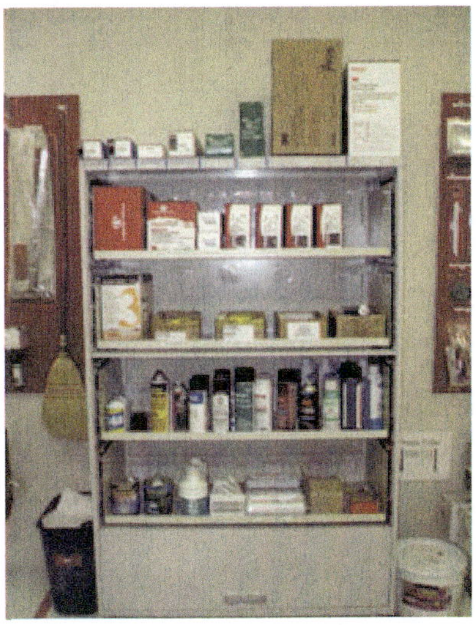

Figure 5.12 Kanban Parts Room

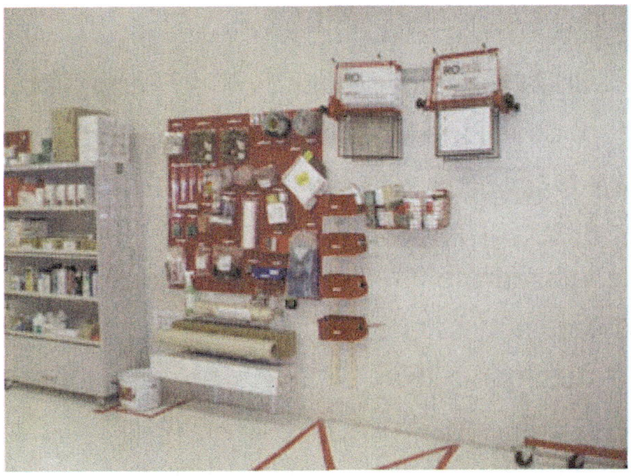

Figure 5.13 Kanban Parts Room

Section Six – Repair Flow

Your body shop is a production line and you must begin to think of it that way. This section will provide some insight on how to establish better flow in your shop. Outside of the teardown process, the rest of the steps include the fabricating and assembling of an automobile, back to its original state: a creative and unique process. I have always been fascinated by the auto collision repair industry, and auto body repair shops were among my first clients when I started Kaizen Assembly. The concept of re-building is extremely innovative and interesting to me.

Changing the mind set of your company is key. Everyone needs to begin thinking about lean production. Now that you have an understanding of the need to be more visual and to develop control over materials and supplies, let's look into making positive changes to your flow.

There are, essentially, five definable processes in auto body repair:

- Teardown
- Prep
- Paint
- Re-Assembly
- Detail (not all shops have this, but they should)

Like a manufacturing or assembly facility, your product goes through different stages of processing to make it complete. Even if you don't have a large facility that allows for physically defined processes, your flow still contains the five steps listed above. George Rybar, of Offutt Collision Repair in Bellevue, NE, was interviewed by a local newspaper about his new facility. He was definitely thinking lean when he designed his repair process. He stated in the article that he designed a linear process and then, put a building around it. He recognized the importance of migrating away from the tradition "bay" style of repairs, where cars stay in one place during the entire process. Instead, his process is linear and it becomes leaner every year. Kaizen Assembly has been working with them for over three years and the linear process was already in place when they became our client. It was a great beginning for their own lean journey.

What is the lean definition of a linear process? Well, in the auto body repair industry, it equates to breaking up these five definable processes into separate work cells, which act as work teams and autonomous departments. Figure 6.1 illustrates this concept.

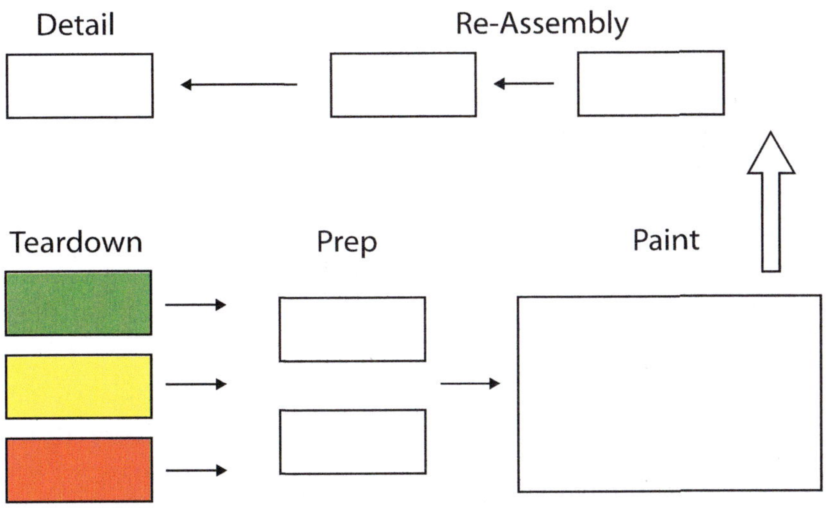

FIGURE 6.1 SHOP FLOW

Each one of the five processes has definable start and end points, which distinguish it from the other processes. Teardown removes the parts, assesses the damage, compares the damage to the estimate, and performs frame repair. Prep prepares the body for painting. Paint is self-explanatory. Re-assembly installs the new or painted parts, including hardware and the addition of fluids. Detail cleans the car, vacuums the inside, and polishes the tires and interior. To establish good flow, you must first identify the unique processes. One of our auto repair clients is in the transmission repair and rebuilding industry. Part of their job is to rebuild transmissions after teardown. Once the assembly is done, the transmission is installed in the car. Call it what you want, but this is a true production process. We have been slowly assisting them into becoming an assembler of transmissions, allowing them to reduce the cycle time of the assembly. No matter how it looks, flow is flow. Once your process has been identified, the flow can be enhanced.

The next step to attaining good flow is applying a time standard to the work. The key to flow is work balance. To illustrate this, let's take a look at the manufacturing industry, once again. Balance is extremely important to any manufacturing process. Every workstation in the flow must have equal amounts of work to attain balance. If the first workstation has 25 minutes worth of work, the second has 10 minutes worth of work, and the third has 32 minutes worth of work, the process, as a whole, is not balanced. The second workstation will be waiting on workstation three and one. Therefore, the next step in acquiring flow would be to conduct time studies on the three workstations in order to determine how the work can be distributed evenly. To further illustrate my point, if this manufacturer was a microwave assembler that produced five different models, we would time the assembly of each model and then, balance the work based on time. This is more difficult to do in the auto body industry...but not impossible.

First of all, in auto body repair, the five processes do NOT have infinite work. Work begins, is completed, and the car moves on to the next process. Here are a few pointers:

Apply historical information to produce your future estimates. Historical information is very valuable, as it illustrates the cycle times previously allocated for the work and the actual hours worked. In lean, we call this flow, "takt" time. Takt is a German word for rhythm. Takt time is the pulse of the process and kaizen experts use it to design most manufacturing processes. Takt time is based on customer demand. Every production or repair process has a pulse, and maintaining flow is directly connected to it. The calculation for takt time is the available time to work, divided by the number of products required to be complete in one day. If there are 420 minutes of work time and the plant needs 10 units completed per day, then takt time is 42 minutes. A unit must be completed every 42 minutes to meet order requirements, so the process is established based on that time.

In the manufacturing industry, this makes sense. Consistent product equals consistent cycle times. In auto body repair, you can apply this with quite a bit of success, even with the complexity and variation that exists in time studies.

Let's suppose that we look at past repairs and the average number of cars completed in a day, and we determine that the total number of car repairs per day should be 5. Our shop has 420 minutes of working time. We divide 420 by 5 (cars) and come up with 84 minutes of takt time. This means that one car's repairs must be completed every 84 minutes, in order to meet demand. Now, with this calculation in mind, let's go through each process in the shop.

Teardown

There are three families of work taking place in the teardown process: simple, moderate, and complex. Simple work is quick, easy, and can be passed on to prep faster. This work includes fixing scratches and dents, or replacing a bumper or other small parts. If no parts need to be ordered, the process is even faster. Moderate work is not as simple, as it requires more parts and more teardown, perhaps even requiring a technician with more experience. Complex work is heavy duty work. It typically includes more parts, frame work, and much more effort throughout the entire process. Cars requiring complex work will be in the shop longer. In regard to establishing flow and time standards, these three levels of work represent the families within the teardown process. Looking at past estimates, can you establish a cycle time based on average repair times for each family? Allocate some time for each level, not exceeding 84 minutes, as a guideline. Although the complex level requires more time, the teardown department must move cars an average of every 84 minutes, as a whole. You must establish a process in which the work and manpower are balanced and allow a car to be completed roughly every 84 minutes.

Prep

In this process, I see two levels of work; simple and moderate. Most likely, complex work would require an entirely new part and therefore, would not apply to prep. For example, a severely damaged bumper would probably need to be replaced, rather than fixed, as repairing would not be cost effective. Also, from a quality perspective, it would make better sense to purchase a new bumper and then paint it and install it...the result would be much better than if you tried to fix the damaged bumper. In our example, the goal is to move the automobiles/work every 84 minutes, so we need to establish a prep area with two locations; one for simple work and one for moderate work. Two locations may mean space for more than two cars, since some prep work may only take a few minutes and the car can pass through quickly, while another car may take longer.

Depending on the insurance company, the prep work may be detailed perfectly in the estimate and if so, you can just use it as your guideline.

Paint

Possible bottlenecks in flow may occur during the paint process but, again, it's all about balance. The set up of the painting operation is the key to reducing cycle time for this process. Time spent preparing the paint guns, mixing the paint, prepping the car, and putting on protective gear is necessary, although it is certainly not value-added activity and therefore, is unavoidable waste inherent to the paint process. However, the minute that paint is applied to the car, the value-added work begins. Implementing 5S into the paint booth area is key to reducing waiting and motion associated with the paint process set up. In truth, the actual painting process is quite fast, when compared to the required prep time. This is always the case, whether you are painting a car, a home, or a room. Proper preparation takes the bulk of the overall cycle time. In regard to our 84 minute example, as long as a car leaves the paint booth within 84 minutes, everything is fine. But, in reality, the drying/curing time can actually be considered a separate process. Is the dry time value-added time? Many debates exist over this, but as far as I am concerned, dry time is required and therefore, must be considered in the overall flow.

Re-Assemble

Depending on the volume of re-assembly, I can see applying the concept of simple, moderate, and complex here, as well. As long as the process, as a whole, can move cars on an average of every 84 minutes, flow can be maintained. Some linear auto body repair processes have established three or four locations for cars in their re-assembly department, and technicians can move into the areas to help relieve any potential bottleneck and keep the automobiles moving. The estimate accounts for re-installation and installation time, so it can be used as a guideline. If you recall, from a prior section, mobile carts have been loaded with the necessary repair parts for each car and are waiting in the parts room. The mobile carts will now be moved to the re-assembly area when the car is ready. Using the numbering system outlined in the last section, the technicians pick the appropriate cart when the associated car enters their area. Remember, the parts room and the mobile carts and numbering system are critical to flow, as they reduce waste of motion and transportation. In the next section, I will discuss visual control and its importance to flow.

Detail

Detail is an operation that exists in some body shops, but not all. There is no time or money allocated in the estimate for cleaning the customer's car. So, in that respect, it would be considered non value-added work. However, detailing can be one of the best marketing and branding tools for the auto body repair industry. Some of our clients have made the detail process the last step in their operational flow. The car is cleaned, polished, waxed, vacuumed, and buffed, to give it a show room appearance. When the customers come to pick up their cars, they notice the appearance first, and often, that's the first compliment they give to the body shop; before they even notice the repairs! If you opt to add this operation to the linear process, takt time must also be taken into account. In this section's example, the detail process must be performed within 84 minutes...and that is likely, doable. Like all departments, make sure that 5S is implemented, to ensure that the detailer does not waste time looking for necessary supplies. When evaluating the families of work in detail, use size as a guideline. For example, cycle time will vary depending on whether you are detailing a 2-door Honda Civic or a 7 passenger mini van. Small, medium, and large families of work are a good model to use for this process; the smaller the car, the shorter the cycle time. Develop a standard time for each level and use it as a guideline to establish the area and streamline the process, so that a car moves every 84 minutes.

Flexing

Flexing is a lean concept that applies to the necessary movement of workers attempting to keep takt time. In lean auto body, it is the movement of technicians to assist others to complete the required work and keep the car moving. A linear process is only as fast as the slowest operation. If the re-assembly area is struggling to adhere to takt time, it doesn't make sense, or add value, for other operations behind them to continue working. At some point, that will only cause wait time or overproduction, while re-assembly continues to be the bottleneck. If re-assembly, for example, is behind, it makes better sense for the upstream processes to stop and flex, in order to help. The flex concept is an important element for maintaining flow and it is very common in most lean operations. The first potential obstacle to flexing, of course, is the adequate cross-training of all the technicians. Ideally, teardown technicians should also be able to perform re-assembly work, and taking it to the next level, teardown technicians should also be able to perform any function in the body shop. At the highest level, every technician should be cross-trained to perform any job in the entire repair process. No matter where your technicians are today, it makes sense to train them to also do the jobs performed next to them in the line of operation. We move to help—that is lean.

Flow is critical to any repair or production process. Establishing your takt time and developing time standards for work, are the two main ingredients. This is followed by being able to shift people and cars to help balance the process. First, identify your key operations. Then, break up the work into families and develop standards for each family. Use these attributes to help balance the line with cars and people (flexing), and you will see a decrease in overall cycle time throughout your facility. 5S will make a huge improvement to flow - you just need to apply the concepts and math to everything in your process.

Section Seven – Visual Control in the Shop

Visual control is a management system used in many lean manufacturing facilities and it is directly applicable to the auto body industry. Like everything I have discussed, the concept of visual control must be customized for the specific industry in which it is to be implemented. Visual control or visual management is a system that is incorporated company-wide, to clearly disseminate information about the process. More importantly, it provides direction regarding what to do, and when to do it, as well as the current status of the operation. Traffic control has its own form of visual management. Stop signs, traffic lights, crosswalks, intersections, highways, and all types of road signs are there to help us maintain control while driving. When appropriate, strategic, visual controls are implemented in a body shop, the results can only be positive.

In this section, I will discuss the following aspects of visual control:

- Visual Layout
- Communication Lights
- Production Control Board
- Material Replenishment

Visual Layout

Section 4 covered the concept and implementation of 5S, as a good starting point. As you now know, 5S is the philosophy of the visual workplace. Often, the concept of 5S and visual controls is interpreted as the same, yet they are very different. 5S is merely the starting point of visual controls and is a part of the visual control philosophy. In regard to visual layout, 5S can be expanded and improved by color coding the various work areas. We will use our five processes of teardown, prep, paint, re-assembly, and detail as examples, designating a color to each area.

Here is the approach:

1. Identify the floor space required for the area
2. Choose a unique color for the area and paint the floor
3. Mark tools and work items with the same color tape, paint, labels, or markers
4. Paint items such as garbage cans, dollies, bins, and racks with the designated color for their area

The actions above are the beginning stages of visual control. If your teardown area is blue and your find a red garbage can in the area, you instantly know that something is amiss. Although this is a small issue, my point is a large one. You can even expand on color coding by distinguishing the three work areas (simple, moderate, and complex) of the teardown area with different colors; red for complex, yellow for moderate, and green for simple. Visual indicators, such as these, will immediately tell anyone in the building, the status of the cars in the teardown area.

Clear identification of aisles and walkways is also important for efficient work, as well as for safety. Visual indicators, such as colors, signs, and direction arrows, can keep people from walking into your work areas, thereby reducing the chance of injury or work interruptions. They also provide visual control for your personnel.

Another element of visual control, in regard to layout, is the clear marking of all KEEP CLEAR areas, such as electrical panels, fire extinguishers, emergency exits, or any other area of your shop which needs to be freely accessible at all times. Develop a visual standard for these areas, such as red hatch marks. You may also wish to position KEEP CLEAR signs in front of each tool board, as a reminder that the tool boards should not be cluttered or disorganized. Image 7.1 is a very good example of 5S, visual control, and keeping clear.

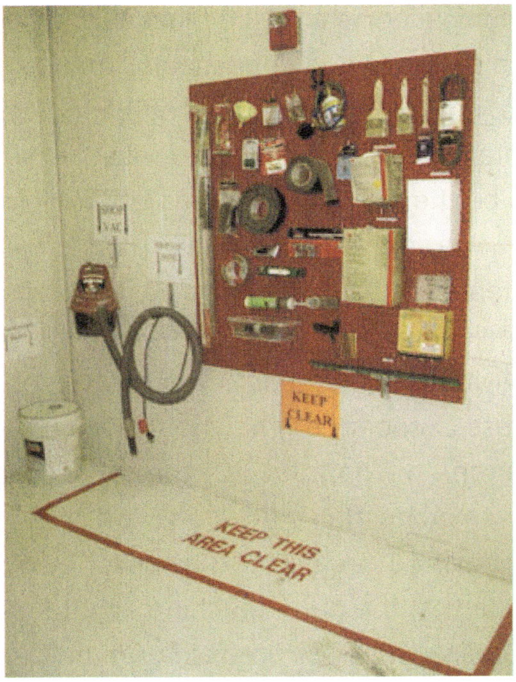

Figure 7.1 Example of 5S, Visual Control, and Keeping it Clear

Communication Lights

Similar in appearance to traffic lights, communication lights should be placed in all work areas. These red, yellow, and green light panels are an effective way that technicians can signal for assistance or for parts, or communicate a problem. The type of communication and/or message you attribute to each light is entirely up to you. However, you must make certain that the signals are clear and distinct and that everyone in the shop is trained to use them and interpret them in the same way. Image 7.2 shows a common product used for this purpose.

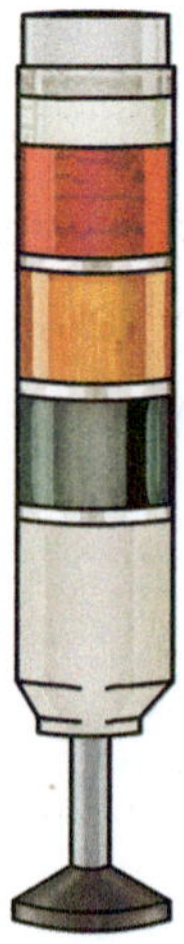

FIGURE 7.2 COMMUNICATION LIGHT

A simple, three color communication light will do most of what you want. However, this type of light also comes with the option of additional lights, sounds, and flashing. Decide what type of signals you wish to implement and then install the lights in your teardown, prep, paint, re-assembly, and detail area, or whatever arrangement is logical for your shop and processes. A red light can be the indicator for a major problem such as parts not fitting, tool breakdown, wrong part, line stoppages, etc. Yellow could mean that a problem is about to occur, such as a tool intermittently malfunctioning. It could also mean that the area is getting low on parts and depletion is imminent. In typical work areas, yellow is used as the proactive communicator, giving warning about a potential problem. Green simply means everything within the process is under control. The visual power of these communication lights is quite amazing. Anyone walking through the shop or simply observing the work area is instantly provided with production status. A yellow light in the teardown area would prompt someone to respond before a real issue has occurred. A red light on the outside of the paint booth may indicate that the paint machine is not functioning and that immediate attention is required. Green lights in all areas inform the shop owner that everything is running smoothly. This type of communication reduces wasted motion and transportation, allowing technicians to signal for help only when required and otherwise, continue their work without interruption. Figure 7.3 illustrates how a light in each area makes a positive impact on the flow of important information.

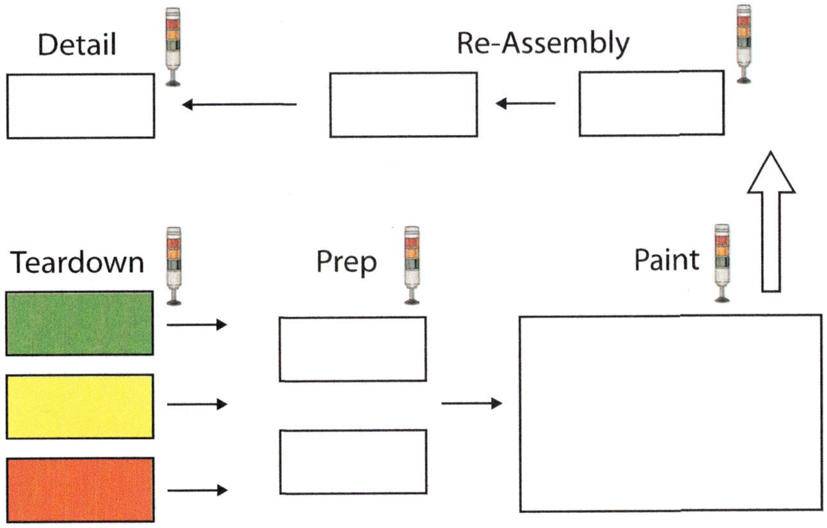

Figure 7.3 Layout Communication

Production Control Board

Imagine yourself as a coach on the football field. Coaching requires constant decision making and the creation of plays during each quarter, which will drive the team to score and ultimately, to win the game. Every game is coached on a per play basis and you want to avoid going into overtime at the end of the game, as that will fatigue the players and may cause decisions that are not part of your playbook. Imagine how difficult it would be to coach a game without a scoreboard to help guide your decisions and your team toward success. Every decision is based on the current situation or status, and with the overall objective of winning. Your body shop is no different. Each day, you and your technicians must make decisions to ensure that the work is done in time for the customer pick up.

Production control boards can be one of the most valuable management tools for increasing output. Of course, in order to be effective, the information on the board must be accurate, but when it is, the results are absolutely AWESOME! A production control board monitors production progress with real-time information, allowing everyone to see if the process is producing the required amount of work. To illustrate this, let's think about an assembly operation, where the line is required to produce 20 widgets per hour and associated takt time has been set to accommodate that. The production control board for this line would look something like Figure 7.4.

Time	Required	Actual	Difference	Current	Comments
9:00	10	10	0	10	N/A
10:00	10	8	-2	18	Tool Broke
11:00	10	11	+1	29	N/A
12:00	10	10	0	39	
↓					

Figure 7.4 Assembly Control Board

As you can see, assembly progress is monitored every hour, and current information is displayed. If the required output is not being met, you can see this, on an hourly basis—not just at the end of the day, when it is too late to do anything about it. Assembly plants have also used takt time to monitor progress by counting how often a widget would have to come off the line; whatever makes sense for each company and process. Each production control board must be designed for the needs of each process—a single template does not exist which would be applicable to any and all processes.

For a body shop, a smart approach would be to use something like Figure 7.5.

Production Control Board (12-15-2009)									
Car	**Estimate**	**Teardown**	**Prep**	**Paint**	**Re-Assemble**	**Detail**	**Complete**	**Due**	**Issues**
Dodge								Monday	
Honda								Monday	Parts
Toyota								Wednesday	
Ford								Thursday	Office
Jeep								Thursday	

Figure 7.5 Body Shop Control Board

This body shop control board works very well for other auto body repair clients, who have found it to be the best management tool in their shops. One shop in Nebraska, displays this board on a projector, so that everyone can view it, update it, and use it as a guideline. It functions much like a scoreboard. Columns include a place to identify the automobiles currently in the shop and those that may still be parked outside. This particular board includes the estimating process, as well as teardown, prep, paint, re-assembly, and detail. It also includes the days of the week, the pick up dates, and a comment column. This particular example is based on the day of the week, in this case, December 15th, 2009. There is a Dodge, a Honda, a Toyota, a Ford, and a Jeep in the "system". Each car has a promised pick up date identified in the due date section. Finally, there is a complete column to indicate when the car is done and a comment section for any issues that should be communicated. Green represents where the car needs to be in the shop, based on the pick up date and the current day of work. Red represents where the car is, in relation to that date, only if it is behind schedule. A car with one green column means that it is on schedule and is currently in the correct process. Two green columns mean that the car is where it should be and, in this case, is actually ahead of schedule. So, in this specific example, the Dodge is on schedule and in the teardown process, as planned. The Honda should be in paint, but it is still in prep, which means it is behind schedule. The Toyota is completed and ahead of schedule, since it should have been in detail.

A production control board is a marvelous visual control tool, which technicians can use to flex workers in and out of work areas, maintaining flow and takt time, as discussed in the previous section. As technicians complete their work in an area, they simply update the board and begin working on the next car. Although one of our clients chose to project this board, that is not necessary. A simple, dry erase board will do nicely for this purpose and meet all of your objectives.

Returning to my football example, as the coach of your shop, you know that you must have a scoreboard that will help you make appropriate decisions and strategies to ensure success and avoid being late for a pick up, while not incurring overtime, which is something all good coaches wish to avoid.

Material Replenishment

The last element of visual control is material replenishment. In Section 5, I discussed the concept of Kanban for material replenishment. The Kanban systems is a visual control approach for effectively handling material and for maintaining the 5S used to organize material and supplies. The concept is important enough to talk about again, and to discuss how it can be applied in your own body shop.

Of all the areas in your shop, the re-assembly area probably has the most parts, supplies, and material. Next in line is the detail area, with all of its soaps, rags, polishes and waxes, followed by paint, then prep, and finally, teardown. After 5S is implemented in these areas, their materials and supplies must be set up on a Kanban system, similar to the one discussed previously. At any one time, the work areas should contain only a minimal amount of material. Developing minimum and maximum levels, as we discussed for the parts room, is critical. Your value-added repair processes are not warehouses! The parts room is to supply the areas with the material, supplies, and part racks when the appropriate signal is given; **not before**. Technicians are to stay focused on the repairs and the automobiles in front of them, as leaving the area to retrieve parts and supplies only slows down the process flow and reduces the focus on quality.

The production control board information should be displayed in the parts room. This gives the parts room employees a visual indication of the cars that will eventually be heading to the re-assembly area and allows them to be proactive, rather than reactive. For example, when the Honda Civic leaves the paint booth and is drying, the technicians mark the update on the board. The parts room worker observes the update and knows to deliver the mobile cart with the parts to the re-assembly lane, placing it in the correct lane according to the corresponding customer number. As a reminder, the parts carts should be assigned the same number as the car that is in need of these parts. Another option would be to implement a light system for the parts carts. When the painter has completed their paint job and the car is drying, they flip a switch, signaling the parts room that the parts cart is now needed in re-assembly.

Visual control of the parts carts reduces the clutter of unnecessary carts in the re-assembly area, helps monitor flow of the parts through the process, and helps control incoming and outgoing parts. It also allows the parts room to prepare carts in the appropriate order, rather than just randomly filling carts and cluttering up the parts room. Why order or prepare parts carts, just to have them sit around and take up space?

The parts room workers are also required to check for incoming Kanban cards in the repair areas, and to replenish as needed, to avoid depletion of all supplies on hand.

Visual control is powerful. Have no illusions or doubts. You will find that your shop will operate at optimum performance levels and with the highest degree of professionalism, making you stand out as an industry leader within your competitive region.

Section Eight – Shop Kaizen

By now, many of you reading this implementation guide may be thinking, how will I get this lean journey going? Where will I find the time and resources to start the actual implementation? Since I need my skilled technicians to support the repair process and I need my office staff to administer paperwork and provide customer service, how will I find the time for lean? How will I manage to keep day-to-day operations going, and still implement the lean philosophies? Good for you! Every company that is considering lean has the very same questions.

This final section will provide implementation advice on how to effectively manage the journey, so that you can find the time and people to support it and achieve the desired results. This section will be dedicated to kaizen and kaizen events.

Kaizen

Kaizen is a Japanese word for continuous improvement. Essentially, kaizen is a philosophy of ongoing improvement that needs to be instilled in the minds of the culture wishing to obtain optimum results in productivity and profit. This philosophy of continual improvement emphasizes the importance of involving everyone in the organization. It assumes that our everyday lives should focus on continuous improvement efforts. This is so obvious and natural in a world class organization, that they often forget that they possess it. Improvements through kaizen are to be integrated into normal day-to-day activities, with the focus of reducing waste, creating standards, and having a clean, organized body shop. Basically, the kaizen philosophy asks that you continually question the current state of work, in order to find a smarter, safer, and better way to do it. That is kaizen.

Kaizen Events

A kaizen event is a bit different from kaizen, itself. Kaizen events are often called rapid improvement projects and are a great mechanism for lean implementation. Kaizen events are the project management side of lean, and they can be conducted in any type of company. Kaizen events have three components: people, time, and the area. Let's make this as simple as possible. If you want to implement 5S in your shop, you will divide/separate the implementation into projects. You will schedule a kaizen event for the re-assembly area, with the team focus being the implementation of 5S. The team will be given a specific time frame; perhaps two days. For those two days, 5S implementation will be the sole focus. At the end of the two days, 5S will have been implemented in the re-assembly area, at least in regard to this manual's guidelines. Simple.

Kaizen events are the team based approached to lean, and therefore they provide for a quick implementation. Now, keep in mind, you are on a lean journey so you will have multiple events every year, depending on time and resources. The idea is to keep improving your process forever. To ensure successful kaizen events, your body shop needs to create its own unique company kaizen program, containing the following attributes:

- Lean Communication
- Kaizen Committee
- Event Planning

Lean Communication

Place a dry erase board in the break room. As lean projects (kaizen events) are scheduled, chart the team, the area, and the dates that the event will be conducted. As the owner, you must communicate that lean is a way of life for the shop. Posting clear information about upcoming kaizen events is critical.

We also recommend a suggestion box or similar system for gathering ideas for future kaizen events. Encourage and promote suggestions from all your technicians. Keeping pertinent lean information current and in everyone's sight, will promote adoption of a lean mentality.

Kaizen Committee

Form a continuous improvement committee that will assist with planning, scheduling, and monitoring kaizen projects. The committee should include office workers, the owner, the front desk assistant, estimators, and employees who work in the parts room. It should also include lead technicians from the body shop, who can represent all the technicians and their interests. The Kaizen Committee will select the goals, team members, the area, and the dates for kaizen events. The committee should meet once a month, even when events are not taking place, to discuss waste reduction opportunities, lean training needs, or anything related to your lean journey.

Event Planning

Kaizen events require solid, up-front planning, as well as follow-up, to ensure success and positive results. Try to follow a 4-week preparation system, which allows for adequate preparation for the events.

4 Weeks before the Kaizen Event

- Select the Area
- Select the Team
- Schedule the Kaizen Event
- Order 5S Supplies
- Establish Goals

2 Weeks before the Kaizen Event

- Verify Team Selection (vacation, light duty)
- Get an Update on Supplies
- Start Looking at the Repair Schedule

1 Week before the Kaizen Event

- Prepare the Shop Floor for the Absence of Certain People
- Place all Supplies in the 5S Supply Box (if it is a 5S event)
- Select an Area for the Team to Meet
- Final Verification of Team Members

This guideline will help ensure that production on repairs is not hindered or compromised during the event. It will also ensure that the team and supplies are ready and that everyone is aware that a lean project is about to begin.

Conclusion

The concepts of lean manufacturing can be applied in the auto body repair business. However, they must be tailored to fit for success. All my clients, both manufacturing and collision repair shops, can vouch for the impressive results that come from lean. The auto body repair industry is highly competitive and involves a lot of different players. This manual was specifically designed for the auto body repair industry, in an attempt to enlighten and provide you with the starting blocks to get your lean journey moving. You will find successes and challenges in your vision for lean, but the end results will be improved quality, increased output, increased productivity, reduced inventory, higher morale, reduced cost, and increased profitability. If these attributes/results interest and excite you, go lean. You won't regret it.

The 8 Wastes

Overproduction: the act of doing more than needed, before it is needed, and faster than necessary

Example:

Extra Processing: the act of redundant effort, extra steps, and working harder than needed

Example:

Motion: movement of technicians, walking and searching for items or people

Example:

Transportation: movement of parts, material, supplies, and cars

Example:

Waiting: the stopping of value-added work (repairs)

Example:

Inventory: excessive supplies, parts, and material

Example:

Defects: repair mistakes, customer returns, and rework hours

Example:

Non-Utilized Talents: not using technicians to their fullest abilities

Example:

5S

Sort: the act of removing and discarding all unnecessary items from the work area

Set-in-Order: the act of organizing everything, so it is easily identifiable in a designated place

Shine: the act of cleaning everything

Standardize: consistency in the application of the visual workplace

Sustain: maintain and improve upon

List of Figures

5S for Auto Body Solution Package

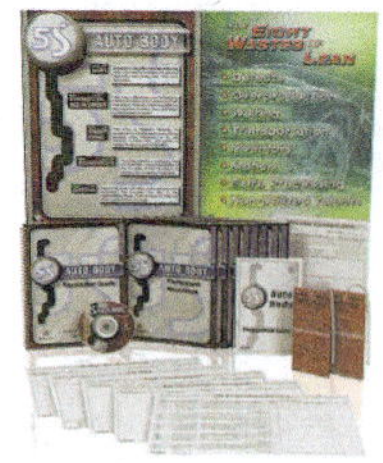

Our 5S for Auto Body Solution Package introduces you and your company to the uses and benefits of 5S. This package is tailored to the Auto Body Industry and provides the context needed to effectively communicate 5S to your employees and managers. We have incorporated years of experience into our training package which provides the greatest leverage point to success. You will be able to turn your business around and put your operations ahead of the competition. By implementing 5S you are setting the stage for greater profit, expansion, and customer satisfaction. 5S will bring your employees together to create a safer, more enjoyable, and more easily navigable work area.
ISBN 978-1-926537-14-6 | 2010 | $549.99 | Item: 25

Introduction to Lean Auto Body Solution Package

The Introduction to Lean Auto Body Solution Package guides you through the elements of Lean and how it applies to Auto Body Repair. You will be presented with a history of lean, the 8 Wastes, and provide a simulation to illustrate and highlight the key elements of Lean for your business. Your employees will have a better understanding of the mindset behind Lean and the 8 Wastes that are inherent to every department within the Auto Body business. With our training package you are able to begin the process of improving your company and achieving superior results.
ISBN 978-1-926537-16-0 | 2010 | $399.99 | Item: 24

5S Training Package

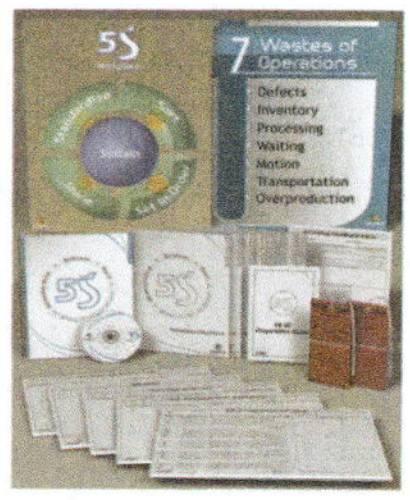

Our 5S Solution Packages will help your company create a sustainable 5S program that will turn your shop floor around and put you ahead of the competition. All of the benefits that come from Lean Manufacturing are built upon a strong foundation of 5S. Enna's solution packages will show you how to implement and sustain an environment of continuous improvement.

Version 1: Sort, Straighten, Sweep, Standardize and Sustain
ISBN 978-0-973750-90-4 | 2005 | $429.99 | Item: 12
Version 2: Sort, Set In Order, Shine, Standardize and Sustain
ISBN 978-1-897363-25-6 | 2006 | $429.99 | Item: 17

Study Mission to Japan

We are excited to present an exclusive trip to the birthplace of Lean. We provide a unique one-week tour at a reasonable, all-inclusive price that will guide you to a better understanding of Lean Manufacturing principles. Enna has exclusive access to Toyota and Toyota suppliers due to our publications of Dr. Shigeo Shingo's classic manuscripts. You will have one-on-one access to Japanese Lean Executives and learn from their experiences and solutions. We also offer custom private tours for executive management teams over 12 people. Join us on our next tour by visiting www.enna.com/japantrip and register online or by telephone at: +1 (360) 306-5369.

Phone: 1+ (360) 306-5369 **Fax:** (905) 481-0756 **Email:** info@enna.com

Other products from Enna

Mistaken Kanbans

Let Mistaken Kanbans be your roadmap to guide you through the steps necessary to implement and successful Kanban System. This book will help you to not only understand the complexities of a Kanban System but gives you the tools necessary, and the guidance through real-life lessons learned, to avoid disastrous consequences related to the improper use of such systems.

ISBN 978-1-926537-10-8 | 2009 | $27.99 | Item: **919**

The Toyota Mindset

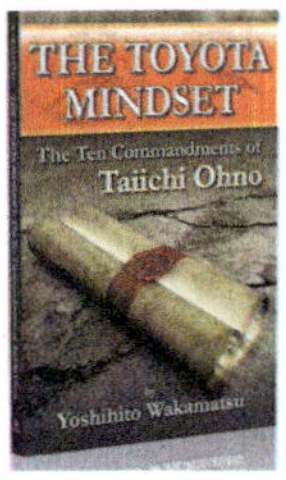

From the brilliant mind of a legend in the LEAN manufacturing world comes the reasoning behind the importance of using your intellect, challenging your workers and why continuous improvement is so important. For anyone who wishes to gain insight into how the Toyota Production System came to be or wants to know more about the person behind TPS this book is a must read!

ISBN 978-1-926537-11-5 | 2009 | $34.99 | Item: **920**

The Toyota Way in Sales and Marketing

Many companies today are trying to implement the ideas and principles of Lean into non-traditional environments, such as service centers, sales organizations and transactional environments. In this book Mr. Ishizaka provides insight on how to apply Lean operational principles and Kaizen to these dynamic and complicated environments.

ISBN 978-1-926537-08-5 | 2009 | $28.99 | Item: **918**

Organizing for Work

When approached from the Lean perspective, H.L. Gantt's Organizing for Work provides a window into the American origins of the 2nd Pillar of Lean – Respect for People. Gantt, the creator of Gantt charts, galvanized the human aspect of efficiency with razor sharp clarity. Production improvements go astray because we have "ignored the human factor and failed to take advantage of the ability and desire of the ordinary man to learn and improve his position."

ISBN 978-1-897363-80-5 | 2007 | $21.99 | Item: **910**

To Order:
Mail orders and checks to:
Enna Products Corporation
ATTN: Order Processing
1602 Carolina Street, Unit B3
Bellingham, WA 98229, USA
Phone: +1 (360) 306-5369 • Fax: (905) 481-0756
Email: info@enna.com

We accept checks and all major credit cards.
Notice: All prices are in US Dollars and are subject to change without notice.

9781926537177